THE NATURE OF FEAR

THE NATURE OF

FEAR

Survival Lessons from the Wild

DANIEL T. BLUMSTEIN

HARVARD UNIVERSITY PRESS

CAMBRIDGE, MASSACHUSETTS

LONDON, ENGLAND

2020

Library of Congress Cataloging-in-Publication Data

Names: Blumstein, Daniel T., author.
Title: The nature of fear : survival lessons from the wild /
Daniel T. Blumstein.
Description: Cambridge, Massachusetts : Harvard University
Press, 2020. | Includes bibliographical references and index.
Identifiers: LCCN 2020005091 | ISBN 9780674916487 (cloth)
Subjects: LCSH: Fear. | Fear in animals. | Emotions
and cognition. | Intuition.
Classification: LCC BF575.F2 B56 2020 | DDC 152.4/6—dc23
LC record available at https://lccn.loc.gov/2020005091

To my students and colleagues,
from whom I have learned so much

Contents

THE NATURE OF FEAR

PROLOGUE

My journey to understand the nature of fear began in 1986, in Kenya. After spending a month studying monkey behavior in Kakamega National Park, a spectacular remnant of the West African rain forests, I took a train to Nairobi, then to Mombasa, and started the first long leg of a cycling trip. My mountain bike was overloaded. A sleeping bag was secured to the top of the rear rack and a tent tied to the top of the front rack. My bike was wobbly and the road was bumpy, so I straddled the line between tarmac and sandy roadside as I cycled toward Tanzania. Unlike the trans-African highway or streets around Nairobi, this road felt reasonably safe. There was little traffic. Periodically, a heavily laden lorry or an occasional bus passed by, belching black diesel exhaust.

Late in the afternoon I was pedaling slowly up a long, straight hill at that interminable speed every cyclist knows—moving, but barely faster than a walk. Ahead I saw three young men squatting on the side of the road. As I approached, I greeted them with an enthusiastic *jambo* ("hello" in Swahili). One of the men was hunched over as I passed, just a few feet away, and I saw out of the side of my eye a glare from another. A split second later a rock the size of a soccer ball hurtled toward my head, thrown by one of the men. I ducked, putting up a shoulder to block the rock. Astonishingly, even though my mountain bike was so heavy, I was not knocked over. But I was hurt and scared. The rock struck a shoulder I'd dislocated only a few weeks prior. With a grimace of pain, I turned to look behind me and saw the men sprinting toward me.

Adrenalin surged through my body. I pedaled furiously. Somehow, I found the strength to cycle up the hill and escape. I kept riding. A few miles later, I heard a truck behind me. Unsure of whether the men would be on the truck, or if someone on the truck might help me, I hesitated—just a moment—before making my decision. I jumped off my bike and stood, gasping, in the middle of the road, waving frantically. Thankfully, the driver and his passenger stopped. I explained what had happened and asked for a ride. They hoisted my bike into the back of the truck and told me that the area was very dangerous. People were frequently attacked by thieves. I was still shaking with after-effects of the adrenaline when they dropped me off after a few miles. After thanking them profusely, I rode toward Tiwi Beach along a sandy track.

The next day, despite the spectacular location, the coral reef I explored, and the delicious fresh fish that I ate, I found it difficult to relax. My thoughts kept returning to the event. Long after the adrenaline response gave my body the energy to pedal away and survive, fear continued to occupy my thoughts. I looked at strangers with a newfound apprehension, worried about cycling alone. I was no longer confident in my ability to estimate the risks I was taking.

While I attempted to tuck this experience away at the back of my mind, I still had many questions about it. Why did it affect me so profoundly—both physically and psychologically? I was genuinely afraid that I was going to be severely beaten or killed. I'd not had such an intense, fearful experience like that before. It was a sharp departure from my day-to-day emotional responses. The surge of adrenaline was overpowering, but it was also empowering; it likely saved my life! What happened to my body when I had this surge? How did this massive physiological response come about? What other sorts of events could elicit it? Why did I keep replaying this traumatic event in my mind? Can events like this trigger recurring nightmares and post-traumatic consequences? How long should I have let this influence my behavior?

To understand why I responded the way I did, but more importantly, to understand why we, as humans, respond to fearful situations, we must understand our evolutionary history. Over millions of years fear has kept our ancestors alive—not only our human ancestors but our nonhuman ancestors too.

My adrenaline response had its origins about 550 million years ago in the evolution of specialized nervous systems in worms. This specialization required a variety of neurochemicals to selectively modulate and coordinate activity, such as the control of responses to life-threatening experiences and situations. Whether balling up to hide or slithering away from a threat, animals evolved the ability to have more complex and, ultimately, conditional escape behavior.

About 250 million years ago, the origin of sociality in termites introduced social stressors that, much like predators, threatened an individual's ability to survive and reproduce. These social stressors included dominant individuals within a social group. The costs of being subordinate involved losing access to food and other important resources. Humans must account for more subtle social stressors in the survival and reproduction equation, like the risk of losing socioeconomic status or status among friends. Social position is related to the ability to access resources to care for yourself and your family. Even though these stressors may seem far removed from escape behavior, evolution works with what it has rather than creating entirely *de novo* traits. When social threats arise, they engage some of the same physiological and psychological systems that previously ensured safety from predators. Thus it is reasonable to assert that it was the evolution of the ability to deal with predatory threats that generated a suite of chemicals and stress responses, as well as fears and anxieties, that enabled our ancestors to survive and reproduce.

In other words, we humans are the descendants of a long line of fearful ancestors who successfully survived predators. Our primate lineage diverged from its tree shrew–like ancestors about 55–65 million years ago, and our most recent ancestors are the chimpanzees and bonobos, from which we diverged about 7 million years ago. Our primate ancestors inherited the physiological abilities to respond appropriately to threats from their ancestors, and those ancestors inherited skills from the ancestors that preceded them. Thus, after generations and generations we have a wonderfully diverse set of antipredator adaptations, including neurochemical, behavioral, physical, and life-history responses. Many of these stem from the beginning of a specialized nervous system. We can view our fears and our responses to fearful situations as part of the great tapestry of life that both preceded us and surrounds us. And, just as we learn from evolutionary history, I believe that we can learn a great deal by observing the life around us today.

For over thirty years I have studied antipredator behavior in the field and in the lab. I have observed a variety of species around the world, from marine invertebrates with simple escape behaviors to birds, lizards, and mammals with a full complement of antipredator adaptations. I have conducted long-term, detailed, and integrative studies of antipredator behavior in marmots—large alpine ground squirrels—and I have studied antipredator behavior in plants. And most recently, with collaborators, I study fearful behavior in humans.

I've noted many similarities between nonhumans and humans, as one would hypothesize based on our shared evolution. As an example, after hearing a clap of thunder or fireworks, a dog or cat visibly shakes or hides. This shaking or hiding behavior also occurs in children. Dogs, cats, and humans share a response from the same neurochemical processes. Fear binds us to our ancestors because it is a mechanism that ensures survival in a risky world. In this book I consider how we can apply some of these observations, supported by our evolutionary knowledge, to better understand animals' and humans' behavior.

When we attribute human characteristics to nonhumans it is called anthropomorphism. It's considered sloppy thinking to assume that nonhumans perceive things exactly as we do. It is important to note that anthropomorphism is also considered taboo among working biologists because it tempts us to infer cognitive abilities to nonhumans that could be accounted for with a parsimonious "stimulus-response" explanation. For instance, is our dog "happy" when I pick up the leash before walking him? Or has he simply learned through simple associative learning that I always put a leash on him immediately before we go out the door on a walk (where he gets to roll in putrid scents)? Does a dog "anticipate" the meal she gets at a particular time of day? Or has stomach acid been secreted to prepare for digestion because her stomach is empty at 7 AM every day? Scientists who invoke anthropomorphism would say that we cannot assert that an animal is happy or sad, able to anticipate its meal, or even fearful in response to a particular event.

I do not assert or believe that animals perceive or process things exactly the same way that we do. But is it really a stretch to assert that animals share a similar emotion—fear—with us? I suggest not. While some may believe *everything* we see in humans is unique, there is much evidence that we share many traits with our ancestors. The primatologist Frans de Waal makes a compelling case that we may be in *anthropodenial,* which he defines as the denial that there are human-like characteristics in animals and that we share animal-like characteristics. Every species is the product of its unique evolutionary history. However, we share many neurophysiological mechanisms with nonhumans, and a variety of social and predatory stimuli trigger identical neurophysiological responses. Accordingly, the social stressors and predatory stressors we'll learn about in the coming chapters often trigger similar responses across taxa.

Importantly, using insights from nonhumans to understand humans is not novel. In science it is commonplace. Many scientists use animals as model systems to better understand developmental, genetic, physiological, physical, and behavioral human pathologies. We study fruit flies and zebrafish, mice and rats, canaries and zebra finches, and rhesus macaques and tamarins. By studying a diversity of animals we can identify multiple ways to solve a particular problem. Or we may learn that there is a single solution so effective that it was passed on through evolutionary time with few modifications. We do so because of our evolutionary affinity with life on earth, as Neil Shubin so aptly noted in his book *Your Inner Fish.* Indeed, it is only through studying our shared evolutionary history that we can make sense of the traits we have today.

However, science progresses by developing, testing, and refuting hypotheses. One should never believe a certain hypothesis; one should test hypotheses and draw conclusions from data. I will review many scientific studies and share the conclusions that the investigators drew from their data. And in some places I will suggest new insights from these studies that may apply to humans. I hope that by the end of our journey together these insights will help you understand why you might react to certain stimuli and messages the way that you do. But, if you wish, consider the insights for humans that emerge from my discussions as evidence-based hypotheses because not all have been formally tested.

From the outset, consider that fear is costly. Foraging in the open, flocks of birds expend valuable energy by quickly taking flight, en masse, when a raptor appears. While they successfully escape the raptor, the birds leave behind their food source. When a shadow passes over a giant clam, it quickly retracts its mantles and closes its shell. By doing so clams protect their valuable tissue, but when closed they are unable to photosynthesize, losing the opportunity to acquire energy and grow. For humans, experiencing alarm—heart-racing, sweat-inducing, eye-opening fear—often results in high costs to personal welfare, health, and productivity. Almost one in five adult Americans suffers from some form of an anxiety disorder each year, and one in four has an anxiety disorder at some point in their lives.

What we learn about the nature of fear may help us better understand ourselves and improve our lives. It can also have profound biomedical implications, as my good friends and colleagues Barbara Natterson-Horowitz and Kathryn Bowers so clearly described in their book *Zoobiquity.* By asking a deceptively simple question, "Do animals get (fill-in-the-blank medical condition)?," novel insights into a variety of human medical conditions emerge from the study of a diverse set of species. For instance, when a predator is detected, a common response is fear bradycardia. Animals freeze, and their heart rates decline. The authors trace this response back to an effective antipredator behavior for fish avoiding shark predation. Since sharks have specialized heartbeat detectors, fish that suppress their heartbeats upon sensing a shark become less detectable. They argue that this ancient response underlies humans' propensity to faint in highly emotional and stressful situations.

Even though we learn about our shared behaviors from studying other species, I am not denying the fact that humans are unique in many ways. Our advanced cognitive abilities are unquestionably unique (thus the concern with uncritical anthropomorphism). For instance, human language has a variety of what the anthropologist John Hockett called "design features." These include things like learnability and semanticity (the ability of signals to refer to things) as well as the ability to have discrete, as opposed to continuously varying, utterances. These abilities are shared with some but certainly not all nonhuman species. And language can be used to lie or deceive. Some but fortunately not most nonhumans have this ability as well. But as far as we know, humans and almost no others communicate about things both in the past and in the future; Hockett called this "displacement." Honey bees seem to be the only other species with this skill; they use their waggle dances to tell hive mates about past experience with food. Such higher-order cognitive abilities—the ability to talk about ourselves, the ability to tell stories, the ability to have abstract reasoning—are seemingly uniquely human.

The extensive literature on human behavior also illustrates many of our cognitive biases. These are wonderfully and artfully summarized in the "Cognitive Bias Codex." For instance, Murphy's Law ("anything that can go wrong will go wrong") is an example of a simplifying cognitive bias. While in some situations assuming the worst is a good idea, Murphy's Law and other cognitive biases may generate erroneous decisions. Thus, by understanding how we assess risk and the biases associated with these assessments, we can empower ourselves to make better decisions. So this means that to better understand our fears, we must also dive into the literature and learn something about human biases associated with risk-taking.

Besides learning about the nature of fear, another goal of this book is to connect you with my scientific colleagues. Personal stories fill these pages, and for that reason, I intentionally use their first names. I also select and highlight relevant stories and projects that emerged from published undergraduate research because I believe that, properly inspired and guided, anyone can make meaningful scientific discoveries.

To begin our journey into terror, we will venture to Pakistan, where an inopportune shoulder dislocation taught me about the neurophysiology of fear and anxiety. We'll also learn about the remarkably consistent stimuli that elicit fear in three different modalities. Dangerous stimuli, it turns out, have predictable sounds, smells, and sights. Many species, including humans, respond reflexively to these. With this knowledge, we'll be better prepared to understand how humans and animals manage predation risks and enhance their abilities to leave descendants.

We'll see how animals learn about threats and consider what that teaches us about human behavior, including post-traumatic stress disorder (PTSD). We'll venture into the surf off the California coast to where my then-ten-year-old son had a bad day surfing—one that has created lasting memories and adaptive behavioral responses. We'll wander through the geography of fear and learn how some wise organisms reduce their exposure to threats. And I'll introduce the economic approach to studying fear that I use in my daily research: all organisms must trade security with the need to acquire other resources. This simple and fundamental rule underlies all of our decisions. Successful organisms must make these trade-offs and live with risk.

In the steaming hot, pre-monsoon forests of India I learned a lesson about living with risks from a peacock's call. Animals communicate about dangers and, by communicating with and listening to others, can increase their chances of survival in a predator-filled environment. But fear and the loss of fear both have consequences. We will venture to my Colorado field site, where human presence has altered the behavior of surrounding predators, and we will venture to Yellowstone National Park, where the elimination of wolves has changed the biological and physical landscape. In both cases, the loss of fear is implicated in these changes. Managing our future biodiversity rests on understanding fear's effects and perhaps embracing the fear stimulated by large predators.

To really understand the impact of fear, a worldwide natural history expedition isn't enough. Fear has profound consequences in many decisions that humans make, both individually and as a society. We will learn lessons for security and defense from a multidisciplinary group of university scholars, policy experts, and warriors. And at the end of our journey we'll combine our new intellectual toolkit with an understanding of some key human biases. I propose fifteen lessons about risk and fear that will help us make better decisions. Why? Because humans are remarkably bad at properly assessing risk. We fear being killed by a shark more than we fear being killed by falling coconuts—even though many more people are killed annually by coconuts than sharks. By identifying and understanding these evolved biases and our susceptibility to responding in certain all-too-predictable ways, we can individually and collectively make better decisions.

Armed with this knowledge, you'll be prepared to interpret fearful and fear-based situations around you. You'll appreciate that it is impossible to live life without risks. You'll realize why fear makes us human, and how, through our behavioral responses to fear, nonhumans and humans are inextricably connected.

Find a nice safe place to read. You're in for a wild ride!

1

A SOPHISTICATED
NEUROCHEMICAL COCKTAIL

Between 1989 and 1993, I spent three to six months a year studying antipredator behavior of golden marmots—large alpine ground squirrels—in the spectacularly beautiful, rugged, and remote Karakoram Mountains of northern Pakistan. The marmots lived in a high, uninhabited meadow called Dhee Sar, which in the local Wakhi language means "above where people are." To get there we started walking at a Khunjerab National Park police check post on the Karakoram Highway at about eleven thousand feet above sea level, crossed a river, meandered up a valley to a seasonal shepherd's camp, then turned left and began an almost three-thousand-foot steep climb to my tented research camp at 14,400 feet above sea level.

On one research expedition we were trekking back from the study site when the trail collapsed beneath my feet. A slip could have led to a long slide down a hill into the rocky river. Thankfully, in my left hand I had a walking stick that was firmly planted in the loose scree slope. Still, with the collapse of the trail, I fell. The sudden, jarring plunge immediately dislocated my shoulder. I'd dislocated my shoulder twice before—once skiing in Colorado and once in a swimming pool in Kenya—but I thought that my shoulder had been properly rehabilitated. Apparently not. Fortunately, we were only a hundred yards or so away from the police and park check post on the Karakoram

Highway where a van could transport me and the volunteer assistants back to civilization. Unfortunately, the check post was across a raging river, and I was in excruciating pain.

I asked my volunteer assistants to position me over a nearby boulder. Then I instructed the group to tie a weighted backpack to my left wrist to fatigue the knot-tight muscles. In time, and in theory, the shoulder would relax and slip back into place. At least that's what I'd read. It was a day's drive to the nearest hospital, so my book knowledge would have to suffice.

In incredible pain, I asked if anyone had Valium to relax my spasming shoulder muscles. Since the research site was so remote, I'd previously asked all members of the expedition to purchase particular medications to combat diarrheal diseases. When they discovered that they could also buy Valium over the counter, most jumped at the opportunity. There was no shortage of Valium. This experience reminded me of the scene in the 1979 comedy *Starting Over,* where recently divorced Phil Potter has a panic attack while shopping for bedroom furniture in a Boston Bloomingdale's. His brother, a psychiatrist, looks at him hyperventilating and asks the gathering crowd, "Anyone got a Valium?" Hands shoot out of pockets and handbags, offering vials of the drug. Volunteers surrounding me offered Valium in a similar way.

Twenty milligrams and a half hour later, I was gloriously relaxed. My shoulder muscles loosened, permitting my humerus bone to slip safely back into its socket. Although my memory of the event is blurry, a large and terrifically strong shepherd carried me across the river. I waited in the van, quiet and numb, while everyone else crossed. The rest of the day passed in a Valium-induced haze. My volunteers described with amazement how my neck muscles were so relaxed that my head bobbed along, seemingly unattached to my body, as we drove over severely rutted roads. Valium and other benzodiazepines relax muscles. They also relax psyches. Benzodiazepines are a class of antianxiety drugs that directly target parts of the brain stimulated when there are impending threats. Valium causes cells to release specific neurochemicals and slows brain activity. Instead of feeling fearful about long-term damage to my shoulder or the rough ride ahead, I was calm.

While Valium enhances the effects of neurochemicals to relax the body and mind, fear does the opposite. Fear creates a highly adaptive neurochemical response that involves specialized parts of the brain, a suite of specialized neural circuits, and a mix of molecules—hormones and other chemicals—that travel between nerve cells, across synapses, and through the blood system to modulate a wide array of responses. In this chapter we'll identify these physiological and neural processes from studies in humans and nonhumans. Familiarity with these mechanisms is necessary if we wish to understand why we fear what we do.

We often perceive fear as an emotional response to threats, largely unaware of physiological responses. Yet, fear stimulates a startle response, changes our heart rate, increases our blood pressure, and dilates our eyes. It changes the way that we assess pain and alters our behavior. These traits are shared with other species, hence, to understand fear we need go back in time to our ancestors. We must go back to our evolutionary roots.

Let's begin about 550 million years ago with a wormlike ancestor and its segmented and specialized nervous system. Before this ancestor only very simple organisms (single-celled organisms) subsisted, either in an active or resting state. Then, once multicellularity evolved, cells developed specialized functions. Our wormlike ancestor's specialized nervous system was a key innovation. Cells could then detect chemicals, or light, or touch, or sounds. Other cells could integrate information from such receptors and, if sufficiently stimulated, fire neurons or release neurochemicals. While selection for such specialized functions was likely associated with acquiring food, there was also selection to identify threatening stimuli and avoid threats.

By the time cartilaginous fishes (sharks and rays) evolved—about 450 million years ago—it is possible to see a more or less anatomically similar autonomic nervous system to that which is shared in vertebrates today. This is important because it means that the sympathetic and parasympathetic nervous systems evolved hundreds of millions of years ago. These complementary networks of nerves, largely out of conscious control, regulate heart rate and respiration rate. In humans, rapid changes in pupil diameter in the eyes (pupillary responses), sweating (recorded using galvanic skin responses), and certain muscle contractions in the face—particularly around the eyes and mouth—are also controlled by the autonomic nervous system. Rapid changes in these can indicate, to perceptive observers, emotional arousal. In fact, lie detectors—the polygraph test (so named because it monitors several physiological processes and traditionally plotted them on graph paper)—are supposed to work by tracking these rapid physiological changes in response to specific questions.

Importantly, the sympathetic nervous system is tasked with the "fight or flight" response. The parasympathetic nervous system reacts to the fight or flight response by returning the body to a "rest and digest" state. Thus, these systems evolved to allow us to effectively function in threatening situations. The sympathetic nervous system reacts and the parasympathetic nervous system creates a homeostatic calming response.

We can also thank evolution for our brain's complex network of activity. When we suddenly encounter a threat, like the attack I experienced on the Kenyan road, a set of our brain circuits becomes very active. Some, like the amygdala (where the nerves that detect threats are concentrated) and prefrontal cortex, are involved in assessing the threat while others, such as the periaqueductal gray, participate in fast fight, flight, and freezing behaviors.

In response to the brain circuits' initial warning (Danger!), a biochemical reaction occurs. Specific genes that make catecholamines (specifically norepinephrine, also known as noradrenaline) begin to copy themselves. Since I do not plan to quiz you later on chemical names, from now on I'll refer to catecholamines more colloquially, as adrenaline. Produced in the adrenal glands and found in both nerves and tissues throughout the body, adrenaline opens airways and eyes while priming muscles for immediate, life-saving action.

You've probably felt a jolt of adrenaline if you've swerved your car and narrowly avoided an automobile accident. Your eyes widened to properly assess the threat or risk. Your nostrils flared to acquire more oxygen to power your muscles. Your muscles tensed in preparation for rapid movement or to protect yourself from impact. Your bloodstream was suddenly filled with glucose, extra energy, to power your escape.

You may feel a rush of adrenaline before public speaking. Or if you've pushed yourself to ski a steeper slope, surf a bigger wave, or climb a rocky ledge where failure could result in injury or death, you've likely had an adrenaline rush. This is what I felt as I pedaled my bicycle up the hill in Kenya, escaping the young men who chased me. My body kicked into a highly evolved defensive mode honed by millions of years of natural selection. Adrenaline is a wonderful thing.

But there are costs to being prepared for life-threatening crises at all times, and our fear responses must be modulated. In humans, the hypothalamus, an almond-sized part of the brain that is found above the roof of our mouths, is directly connected to the pea-sized pituitary gland that rests atop it. Adrenal glands are located above the kidneys. These three actors work together to regulate energy allocation on a longer time scale and collectively maintain and modulate the fight or flight response. After perceiving a threat on the Kenyan road, while an initial burst of adrenaline coursed through my brain and veins, my hypothalamus started to produce corticotropin-releasing hormone (CRH). CRH immediately stimulated genes in my pituitary gland to produce adrenocorticotropic hormone. ACTH traveled through my bloodstream (quickly because my heart was pounding to get oxygen to my muscles) to my adrenal glands, whereupon the hormone triggered these glands to produce corticosteroid hormones from cholesterol.

Cortisol and corticosterone are the major stress-induced corticosteroid hormones in vertebrates. Which one is more common varies by taxa, but interestingly, CRH also directly increases anxiety by activating specific cells in the brain located near the cells that are sensitive to adrenalin. These corticosteroids, like many hormones, have an array of effects throughout the body. Corticosteroids reduce energy allocation to maintenance activities such as growth and reproduction. Their release sends glucose to the muscles to facilitate fighting and fleeing. When an individual experiences an immediate threat, like my experience while bicycling, corticosteroids suppress costly immunological responses, including protective inflammatory reactions. Circulating stress hormones focus an individual's attention on the threat, and other activities and disturbances become secondary since the goal is survival. Recent work has also shown that corticosteroids can be produced in other tissues (including brain tissues), where they have direct and immediate effects generally associated with preparing us for flight. In summary, the HPA axis (the integrated set of the hypothalamus, pituitary gland, and adrenal glands) provides support to the immediate release of catecholamines following the perception of threat. I believe these biochemical responses saved my life.

Brain imaging studies show that there are specific fear circuits—sets of connected neurons—in the brain, and the hypothalamus and amygdala are right in the middle of it all. By using functional magnetic resonance imaging (fMRI) techniques, researchers are able to determine with great specificity the areas of the brain that become activated in response to specific types of stimulation. Activated areas have increased oxygenated blood flow, and oxygenated blood has a different magnetic resonance than nonoxygenated blood. Oxygenated and thus activated neurons can be identified by fMRI. By scanning people as they solve problems or are exposed to various stimuli, fMRI allows the identification of the specific parts of the brain involved in these processes.

While it is helpful to see what areas of the brain are stimulated, some caution must be exercised in fMRI analysis. First, potential statistical errors can emerge from the magnitude of data produced. Researchers search for specific sections of the brain showing activation, but because of the fine spatial resolution of fMRIs, they must compare resonance patterns with many other portions of the brain to rule out clusters of false positives and determine which portions are activated by the stimulus. Thus a statistical adjustment must be made to properly interpret fMRI results, and many early studies did not do this. It took a study showing that a dead salmon had emotional responses to photos to show the need for improved analysis. A second limitation of fMRI interpretation is that it does not have the spatial resolution to focus on specific brain cells. And because similar parts of the brain are activated by a number of stimuli, critics of fMRI studies note that the technique is not necessarily a simple or direct key to understanding brain function. Nevertheless, when properly analyzed and interpreted, the corpus of fMRI studies provides intriguing insights into how we respond to fearful situations.

In human experimental trials, when fear of injury or death is present but not imminent, there is increased neural activity in a part of the brain called the ventromedial prefrontal cortex. By contrast, when threat of injury or death is imminent—as I thought it was when I was attacked on my bicycle—there is an increase in the activity of a part of the midbrain called the periaqueductal gray (PAG). Electrical stimulation of the PAG in nonhuman experiments leads to immediate defensive behavior, while killing or removing cells eliminates fearful responses. I'd suspect my PAG was working overtime on that Kenyan road.

Our current understanding of these fear circuits is that threats are processed along a continuum of urgency. Anxiety-producing stimuli activate more recently evolved parts of the brain in the prefrontal cortex, whereas fear-producing stimuli and situations activate older parts of the midbrain. The amygdala is associated with avoidance and preparation for escape, whereas the hypothalamus and PAG is associated with escape.

Fear and anxiety are neurochemical responses to threats. Glucocorticoids—stress hormones—are associated with the production of alarm calls in the yellow-bellied marmots I study. Females are more likely to emit alarm calls when approached when they have higher levels of these hormones in their bodies. This is a correlative result, but in an experimental study, rhesus monkeys given metyrapone (a chemical that reduces cortisol levels) were less likely to emit alarm calls in a threatening situation. The monkeys that did emit alarm calls did so at slower rates, consistent with a lower level of perceived risk. Thus, neurochemicals can directly reduce fear. This was made clear in my experience with Valium!

But modulating chemicals, whether by medication or not, is costly for the organism, similar to the up-and-down cycle of the fight-or-flight response. Being constantly vigilant and prepared for flight is energetically costly, thus biological systems that up-regulate physiological responses have homeostatic mechanisms to down-regulate the processes. Excessive vigilance is a lost opportunity cost; it may prevent individuals from engaging in other important activities. High levels of glucocorticoids have immunosuppressive effects that make individuals vulnerable to parasites and pathogens. Yet, it's essential to survive the immediate threat in order to reproduce, grow, or fight off an infection.

Thus, a more nuanced way to view the HPA axis, and indeed cortisol and other stress hormones, is that they work together to ensure that animals are allocating energy in ways that promote their long-term survival. When there are few immediate threats, growth, immunological defenses, and reproduction are favored. When faced with impending threats, individuals engage an emergency life history strategy modulated by the HPA axis to guarantee immediate survival.

Beyond a threat to one's life, other stressful events can trigger activation of the HPA axis. Social stressors are notable, both in humans and nonhumans. For example, corticosteroids increase after an animal loses a fight or is glared at by a dominant individual. For some (but not all) species, social stress comes from not being in control of a social situation, as seen in the glucocorticoid levels of subordinate individuals. And this lack of control is costly. These stressors are not only restricted to adults; stress can begin in the womb (or egg) and occur soon after birth.

When threatened by predators, mammalian mothers have heightened HPA activity. The hormones surging through their bloodstream may affect their fetuses or their nursing offspring and may lead to long-term adaptive responses. If mothers are routinely exposed to predatory threats, then their offspring will most likely be born into a relatively risky environment. Life history theory focuses on energy allocation decisions throughout an animal's life. Those who are more likely to die young should reproduce at an earlier age.

There is evidence that maternal programming, modulated through HPA axis activation, is associated with offspring life histories. Specifically, offspring from predator-stressed mothers reproduce earlier or have more offspring because they will likely live shorter lives. But life is all about trade-offs that are driven by scarce resources. By producing more offspring, less energy and care can be invested in each of them. Additionally, after reproducing, exhausted mothers may increase their likelihood of being killed through risky foraging to acquire needed energy. If mothers that follow such rules leave relatively more surviving offspring than females following different rules and these rules have some genetic basis, then the logic of natural selection dictates that they will have relatively higher fitness and the successful rule will be selected.

Mere exposure to predators, or predatory cues, can increase production of corticosteroids and shift individuals into survival mode, directing energy away from growth and reproduction and toward defense. In an elegant series of experiments with snowshoe hares, Michael Sheriff and his colleagues Charlie Krebs and Rudy Boonstra studied how hares respond to predatory stressors. Snowshoe hare populations increase and decrease over time, driven by the population of their main predator—lynx. As lynx populations increase, hare populations decrease until there are not enough hares to feed the lynx, resulting in a dramatic crash of the lynx population. Charlie, now a distinguished professor, has spent much of his professional life studying the details of this highly dynamic population cycle.

By tracking lynx and hare populations, Michael and his colleagues found that hares had higher stress hormone levels at the peak of the lynx population cycle. The mere proximity of a dog walking by an enclosure containing pregnant female hares resulted in higher stress hormone levels in the dam and her offspring. Importantly, stressed dams produced fewer offspring. In a follow-up study, it was shown that the effects of stress hormones on influencing reproductive success persisted in the offspring. Specifically, the offspring of stressed mothers were themselves more stressed. These results provide one of the best examples of stressful maternal programming in the wild.

By exposing pregnant animals to live predators or predatory cues—such as their scents—a number of studies have documented major shifts in the behavior, the morphology, and ultimately in the health of offspring. When exposed to polecats during gestation, female root voles produced sons with elevated stress hormone levels. The offspring of three-spined sticklebacks exposed to predator-scented water swam together in tighter shoals, an antipredator defense. And when exposed to cats during gestation, rat offspring were more susceptible to a chemical-induced seizure than offspring from rats simply immobilized during gestation. Yellow-eyed penguins exposed to human tourists have increased stress hormone levels. This physiological stress interferes with their reproduction. Fortunately, Galapagos marine iguanas routinely exposed to tourists have reduced stress-induced responses, showing that over time, benign threats may eliminate this costly response.

Another type of stress-induced response is broken-heart syndrome, or takotsubo cardiomyopathy. In *Zoobiquity*, Barbara Natterson-Horowitz and Kathryn Bowers describe how this stress-related heart condition occurs in humans when a surge of catecholamines are produced in response to a sudden, life-threatening, or highly emotional shock. Sometimes these surges damage the heart muscles and create characteristic scarring that results in the heart misfiring. Veterinarians identified catecholamine-related muscle damage to the heart and other flight-related muscles in a variety of species of wild birds and mammals following a sudden capture-related death. Veterinarians called this syndrome "capture myopathy." But my own experiences with closely related species puzzled me. When I worked in Australia with captive wallabies and kangaroos, I learned that while tammar wallabies were quite robust and fared well in captivity, parma wallabies (their closest living relative) were extremely sensitive to capture and external stressors, and susceptible to capture myopathy. Why? What explains these differences?

Inspired by these questions, Barbara and I studied the evolution of the vulnerability to capture-related sudden death. We focused on ungulates (hooved mammals, including deer, sheep, goats, and cows) because as species well-studied by veterinarians there was a breadth of literature to analyze. While surveying the literature for reports of capture myopathy or sudden death following capture, we hypothesized the vulnerability would be associated with life-history traits and adaptations to reduce predation risk and hence enhance longevity. In a formal comparative analysis, we found that species with reports of capture-associated sudden death were more likely to run quickly, be social, have larger brains, and live longer. These results suggest that the propensity to be scared to death is likely to be associated with the characteristics identified with living a long life. If we follow this line of reasoning, long-lived species, including humans, may be more likely to be scared to death due to the very suite of adaptations that enable survival. We should thus view capture myopathy not as an adaptation for longevity but rather as a constraint that is a byproduct of selection for other factors that are associated with longevity. Indeed, a broader lesson from this study is that not all traits we see are adaptive.

Some degree of fear and anxiety is ancestral and often highly adaptive. A suite of well-choreographed neurochemical responses inevitably follow a threat or an exposure to a scary stimulus. These fight-or-flight responses are adaptive in that they increase the chances of survival. But even adaptive traits can go awry, particularly when we are in ecologically or evolutionarily novel situations.

We see this in aggression between members of one's own species, an ancestral trait that explicitly involves our fear systems. In many situations, an individual that fights will acquire needed resources, like food, shelter, or access to mates, while the loser will not. Indeed, the fight-or-flight response is not simply defensive, it is offensive. When fighters psych themselves up for a fight, or athletes prepare for a contest, they are activating their sympathetic nervous system. The body prepares for action. Those that are the most prepared may have an edge in a contest. But sometimes these same adaptive responses can go very wrong.

I recently stumbled into a potentially threatening situation near one of the Colorado sub-alpine meadows where I study marmots throughout their lives. We live-trap, mark, and release them and then observe their behavior from afar. We set a lot of traps each day and check them regularly since our goal is to avoid harming the animals. While we were out trapping, three police wearing body armor and carrying automatic rifles were setting up an ambush for an individual with a history of firearm violations who possessed an unregistered gun. They were checking their rifles and preparing a roadblock on a small, private dirt road. I startled them as I cycled up the road. The lead officer, voice tense and eyes wide open, asked why I was there and instructed me to leave. But I had set marmot traps two hours before and it was warming up quickly. Since marmots are adapted to the cold, not the heat, being left in a trap in the direct sun can kill a marmot in as little as thirty minutes. I politely requested permission to proceed. The first officer, rifle in hand, nervously shouted at me about the dangerous situation and instructed me to leave immediately. Nervous police with big guns is a toxic cocktail. If he was a rural police officer without advanced training needed to manage such encounters, he would have reason to be quite anxious. Hours later I received permission to access my traps via another route. I found one marmot who, while warm, was not harmed, and sat quietly in the trap until I released him. With the marmot tragedy averted, I relaxed. However, neurochemicals were likely still surging through the police who, as the news stated, were successfully trapping their subject.

In these kinds of scenarios, where we come into contact with unanticipated threats, we are still controlled by ancestral parts of our brain. We evolved to fight with our hands, not with the incredible power of modern assault weapons or semiautomatic, multiround handguns. This evolutionary mismatch has become a safety problem, a public health problem, a racial problem, and a deeply personal problem.

Training and awareness of our evolutionary response may help us change our behaviors. When threatened, our sympathetic nervous system is immediately activated, and a cascade of neurochemicals prepares us for defense. Before succumbing to this powerful preparation, we must first try to de-escalate and allow the homeostatic processes of our parasympathetic nervous systems to take over. We all need to slow down and think before reacting to fearful or stimulating events. This pause is made more difficult because the biochemical responses of the HPA axis are geared toward action, not thought. But effective training can allow us to gain control of our immediate physiological responses. By taking a moment to think in the midst of a threatening situation, we can make better decisions.

In every stage of life, in every situation, all living organisms, including us, have to make short- and long-term decisions about how to allocate energy to defense or to growth and reproduction. Some of these decisions are conscious, but some are not. Natural selection has favored individuals' actions that permitted them to survive, reproduce, and leave descendants. Our highly adaptive neurochemical response to fear evolved to increase survival. These days, however, sometimes taking a deep breath and pausing before action is essential for survival.

2

BEWARE OF LOOMING OBJECTS

One rainy Sunday afternoon on the UC Davis campus I rescued a young crow. Leaving the library where I'd spent the afternoon reading, I heard a cacophony of crows. They were mobbing a cat who was solely focused on a mostly helpless fledgling, a young crow that left its nest a little too early and was unable to fly. With no one around to assist, I braved the agitated crows, chased away the feral cat, and wrapped the shivering baby crow in my jacket. The mob of angry crows followed me to the office, cawing loudly. I waited until after sunset before I felt it was safe to leave my building. (I've seen Alfred Hitchcock's *The Birds*.) My plan was to climb to the nest in the morning and release the crow where I found him.

Judy, my professor, and my office mate Perri met the red-mouthed nestling the next morning, and we decided it was best to keep him inside for a while. Thus began the karma-killing process of feeding a fledgling bird: we purchased live crickets by the case, dipped them in vitamin powder, and shoved them down the hungry fledgling's throat. It was only weeks later we realized that Crow! (his name) enjoyed tortellini. From that day on the bird ate better than our lab's graduate students. He enjoyed eating many foods, from Italian cuisine to dog food to burritos with salsa. But the salsa made him salivate profusely and puff up some of his head feathers. He'd shake his beak to clean off the saliva, which usually ended up on the burrito and my face.

We let Crow! out of his cage when we were in the lab so he could properly exercise his wings. Unfortunately, Crow! demanded constant attention. One day he was sitting on Judy's chair while she was writing. Casting an eye up at her, he craned his neck and pecked the red power button on her tower computer. BOOM! That got a response! Judy scolded him. Now she had to reboot her computer. Crow! learned that this was a great way to get attention. He began to sit on Judy's chair and gaze at the red button so Judy would play with him. (This only worked for a short time. Judy eventually covered the button with cardboard and tape.) Later, Crow! cached some of his dog food in a disk drive.

Aside from learning about Crow!'s intelligence and social interests, we learned about Crow!'s fears. The janitor would come into our lab every afternoon with a broom to sweep the floor. Crow! found this terrifying. He would fly to the top of the lights and emit very scared cries. Perri thought that Crow! might be responding to the broom pole. To test the hypothesis, she brought in tomato stakes from her home. Bingo! Crow! avoided tomato stakes around the office, even computers with tomato stakes artfully arranged over them. But why would a crow be afraid of a long stick? Possibly because it looked like a snake?

Classical ethological studies show that many species have ingrained fears of certain object shapes or objects that behave in certain ways. By identifying these shapes and behaviors, we understand the sights that elicit fear.

In this chapter we explore what and why certain objects and visual stimuli are generally fear inducing and the consequences of this fear. We learn why animals and humans are prepared to learn to fear certain visual objects but not others. And we learn how animals assess the risk of approaching objects and how this knowledge may help us reduce the incidence of vehicles (including planes) hitting them.

The field of ethology, the naturalistic study of animal behavior, has its roots in Europe. Ethologists interested in animal behavior spend time outside getting their boots muddy while studying different species. By contrast, the field of comparative psychology, based in North

America, studies the minds of rats, pigeons, and some primates in captivity. Ethologists are not averse to keeping captive animals for research, but when they do so they typically focus on nontraditional animals. For instance, Nobel Laureate Konrad Lorenz hand-reared geese to understand (among other things) the ways that they learned who their mothers were (a topic referred to as filial imprinting) and who would make a suitable mate (known as sexual imprinting). Some scientists study animals in nature habitats and in captivity. Eberhard Curio was one such scientist. A vibrant and highly distinguished German octogenarian ethologist, Eberhard in his youth conducted pioneering work identifying the visual cues animals used to distinguish predators. Combining focused studies of both free-living and captive pied flycatchers, a small European bird, Eberhard scared birds with taxidermically stuffed and model predators. He capitalized on the flycatcher's mobbing response.

What's mobbing? Have you ever seen a small bird loudly chasing a crow or raven away from its nest? Or crows chasing a man attempting to save a fledgling? Or, as my wife, Janice, and I witnessed one day while camping beneath a colony of Pacific crows on Washington State's Olympic Peninsula, a murder of crows loudly pecking at a raccoon as it tried to steal eggs and nestlings from the crows' nests? Mobbing is an unmistakable behavior due to the loud, attention-getting vocalizations used to recruit others and to help dissuade potential predators. Mobbing is cooperation at its best.

Because mobbing is easy to identify and the vocalizations are simple to count, Eberhard Curio used vocalization rate as an unambiguous assay of fear. He was one of the first scientists to note that the same predator elicited different mobbing responses at different stages of the nesting cycle—an economic insight we will discuss in more detail in Chapter 6. Flycatchers were well suited for these experiments because they readily nest in nest boxes—wooden boxes with a small hole in them—affixed to trees in the European forests where Eberhard worked. Rather than spending weeks finding nests, Eberhard arranged the nest boxes, and prospective subjects came to him! He started his study by placing either stuffed redback shrikes or models of redback shrikes, a nemesis to nesting flycatchers, outside flycatcher nest boxes. He found that models of shrikes were quite effective at eliciting mobbing behavior. However, when he placed the models upside down, very few flycatchers showed mobbing behavior in response. Curious to identify what it was about the shrike that evoked fear in the flycatchers, Eberhard performed additional experiments.

Redback shrikes have a distinctive black eye stripe that looks much like the Lone Ranger's mask. Eberhard found that intact models elicited a lot of mobbing, but models with the stripe removed elicited very little mobbing. If he made the model completely white or black, or if he removed the eye stripe, there was little mobbing. It wasn't the contrast per se or the presence of black that elicited mobbing by the flycatchers. Eberhard went a step further. He shrank the size of the shrike but retained its coloration patterns. This evoked slightly less mobbing, as one might expect. (Smaller predators should be less of a threat than larger predators.) If he painted black stripes where the eye stripe should be on sticks the length of shrikes, there was no response. By changing the orientation of the stripe to vertical, rather than horizontal, he eliminated the response. When he shaded out the stripe systematically, Eberhard abolished the response. He was onto something.

Eberhard then studied the flycatchers' mobbing response to model pygmy owls—another flycatcher predator. Through these experiments he found that only an upright stuffed owl, not a wooden model or a wooden model with feathers glued on, elicited the mobbing response. By manipulating the number of eyes on the stuffed owls, he found that two eyes, even if they were slightly different colors, evoked mobbing, while stuffed owls with either one or zero eyes did not. The eyes have it! Flycatchers pay attention to their predator's eyes, and it's the presence of eyes that elicits mobbing. It turned out that the black eye stripe masked the predator's eyes, making it potentially difficult to see where the predator was looking and thus making the flycatcher less certain about its personal security. This uncertainty elicited mobbing, which functioned to drive the predator out of the area.

Eberhard Curio then conducted more experiments, placing eyes that contrasted with the eye stripe on the redback models. Bingo: models with eyes scared the flycatchers, who in response to seeing them, mobbed the models. Eyes are important! Further experiments showed eyes had to be in the correct position, as did the eye stripe. Because he could manipulate the models, an asset in his experiments, Eberhard then made owls look like shrikes. The most effective model owls had real feathers. Interestingly, the precise plumage coloration was more important for the owl models than for the shrike models. What about beaks? Add a long Pinocchio-like beak to a shrike model, and the flycatchers still mob it. Similar beaks added to owls elicited less of a response. Placing a red ball behind the head of the most evocative models did not influence the mobbing response. Eberhard continued to conduct more experiments to assess why a predator's eyes evoked so much fear.

Many other species don't like eyes peering at them either. If you're lucky enough to spend time with gorillas in the wild (still on my bucket list), you will be instructed to not look at them directly, lest you upset them. And if you stare too intently at a male baboon raiding your camp while on safari, you'll be on the receiving end of a nasty, canine-revealing grimace. Don't be deceived by the smile—it's a threat—and baboons, with their very large canine teeth and enormous strength, are not animals you want to fight. When a baboon smiles at you, listen to your HPA axis and quietly move away!

Joanna Burger, a biologist at Rutgers University, and her colleagues followed in Eberhard Curio's footsteps by conducting experiments on basking black iguanas. They approached the iguanas wearing half-masks with either small or large eyes. Sure enough, the iguanas fled, though they ran farther away when the approaching experimenter wore a mask with large eyes. But it's got to be more than simply eye-spots or big eyes that elicit fearful responses. It has to be whatever looks scary.

Snakes, it turns out, are scary to many species. That was the case with Crow! But is this fear innate in the sense that animals respond fearfully the first time they are exposed to them? While I had some suspicions based on Crow!'s behavior, captive studies are essential for finding answers to these sorts of questions.

Japanese macaques reared in the laboratory, in complete isolation from snakes, responded to pictures of snakes more quickly than they responded to pictures of flowers. More recent work on Japanese macaques identified the specific neurons that fired only when stimulated by snake images. Such a neural response permits rapid snake identification. But rhesus macaques also hone their predator recognition abilities with experience. Classic studies show that captive, snake-naïve monkeys learn to respond fearfully to snakes by watching the video-recorded behavior of other monkeys responding fearfully to live snakes. They can also learn, by watching the behavioral response of video-taped monkeys, to respond fearfully to toy snakes. But they do not develop a fear of plastic flowers, despite editing of the videotape to show monkeys being terrified of plastic flowers.

What is it about snakes? Humans have feared and vilified snakes at least since writing the Book of Genesis. Carl Sagan speculated that because mammals evolved with mammal-eating reptiles, we have a deep-seated fear of them. Indeed, studies have shown that many people list snakes as one of their greatest fears. Humans respond more quickly and orient their eyes more quickly to images of snakes (and lions) than they do to lizards or an African antelope. Humans are better at detecting camouflaged and degraded images of snakes than they are at detecting nonthreatening birds, cats, or fish.

Spider sightings also generate fear in many humans. My colleague Dean Mobbs conducted an experiment using functional MRI machines to monitor brain activity. Subjects placed one of their feet in an apparatus with boxes placed at different distances from it. Via a video feed, subjects watched as a live tarantula was placed in a box near their feet. As the tarantula was moved closer to the subjects' feet, Dean and his colleagues found increased neural activity in the periaqueductal gray (PAG) and amygdala, which, as we learned in Chapter 1, are the parts of the brain activated by fearful responses. As the tarantula was moved away from the subjects' feet, different parts of the brain became activated (the orbitofrontal cortex), suggesting that there is also a unique set of neural circuits associated with emitting safety signals. Perhaps this is similar to the balancing of neurochemicals to bring the body back to a state of calm after a boost of adrenaline.

How do fears develop simply from seeing something? Do we have an innate ability to recognize spiders? Are we born with arachnophobia? In an experiment with five-month-old human infants, David Rakison and Jaime Derringer counted how long the infants looked at each presented image. (You have to question who puts their kids in studies like these.) The images included a black and white schematic drawing of a spider with spindly legs. They also showed the infants a drawing with the same spider body but with the legs repositioned so it did not look like a live spider; rather, the drawing looked more like a squished spider. The third drawing retained the same body parts and legs, but completely scrambled the parts. This one did not look like a spider at all. The researchers thought by observing how infants, presumably with no experience with spiders, responded by looking to these visual stimuli, we would understand whether humans have an innate ability to recognize spiders. Infants looked at the schematic, the more realistic-looking spider, more often than the other images.

To see if this was because infants formed a concept of "spider," the researchers used a habituation-recovery protocol, which is a common technique employed to study categorization in preverbal children. They showed their subjects photographs of real spiders until they no longer responded by looking at them. At this point the infants were said to have habituated to the spider pictures. Then, researchers showed the infants the same three drawings again. If the babies responded differently to one or more of these images, they were said to have dishabituated, meaning that their response to the real spiders, which had become weaker through habituation, had returned to its full strength. The researchers would infer that the babies perceived the new images differently from the images they had habituated to. The babies responded more to the images that weren't spiders! This result showed that the infants learned to ignore the real spiders and then, by not responding to the schematic spider image, classified the schematic as a spider. In other words, the schematic wasn't classified as different from the photographs, and didn't warrant another look. By contrast, the other drawings were sufficiently different from the no-longer-interesting spider photograph to capture the infants' attention. And this means that they were classified by the infants as different.

To see how spider-specific this was, they also conducted the same experiment with three drawings of a flower: first a fine schematic, then one with features moved around a bit, and finally one that was completely scrambled so it did not look like a flower anymore. They found that, unlike with the spider images, the infants had the same response to all of them. Analyzed together, these results suggest that human infants have an innate ability to recognize spiders. Researchers hypothesized that humans have an inner template that permits us to recognize the spiders.

Behavioral biologists use the word "template" to indicate that an animal has some sort of a reference standard that is used to help it identify or categorize something. Sensory neurons fire when they are roused by stimuli to which they are sensitive. Thus, there are pressure-sensitive, color-sensitive, odor-sensitive, motion-sensitive, movement-sensitive, taste-sensitive, and even orientation-sensitive neurons. One can imagine that evolution has selected for arrays of these sensory neurons to be tuned so as to recognize rather complex visual (or tactile, or olfactory, or acoustic) stimuli. Indeed, there are sets of neurons that fire when they detect an upright but not an inverted face—recall how flycatchers responded to inverted shrike models.

But these templates need not function at the level of sensory neurons. They could be more cognitive and be based on experiences. We could learn to fear something because we associate its presence with an aversive experience, and, in the lingo of psychologists, generalize this specific memory to a more general class of stimuli. As an example, when a soldier is exposed to a violent act, associations may be formed with stimuli related to the violent act. If an improvised explosive device is hidden in a trash can, trash cans may become a fear-inducing stimulus. We will talk more about post-traumatic stress disorder (PTSD) in future chapters, but for now, let's return to how templates could be used for predator recognition in Australian marsupials.

Tammar wallabies—cat sized kangaroos—are nocturnal. In their natural environment tammars spend their days squatting in dense vegetative cover and their nights hopping out to meadows to feed. Their feet, like those of other native Australian grazers, are padded, and they hop quietly along trails, called stamps, in the dark. When foraging, they often congregate in large groups.

Tammars were perfect for our research on fear because the tammars we first studied had evolved on an island, Kangaroo Island, where they never were exposed to foxes or dingoes. Indeed, because Kangaroo Island was cut off from the mainland thousands of years ago, they had no exposure to any terrestrial mammalian predators for about 9,000 years. Instead, their main predator was the magnificent and large wedge-tailed eagle. From Kangaroo Island the tammars were brought to outdoor enclosures at the Macquarie University Fauna Park in Sydney and bred in captivity.

Andrea Griffin was a graduate student when I collaborated with her and my postdoctoral mentor, Chris Evans, on studies of predator recognition of tammar wallabies. I focused on a variety of questions, among them whether tammars had an innate ability to recognize predators and how long animals could retain their fearful responses after isolation from their predators. Andrea wished to see if tammars could be trained to respond fearfully to red foxes. Both Andrea and I shared an interest in knowing whether tammars would have a fear of red foxes even if they had no previous experience with those predators, and particularly if there was no opportunity for them to have evolved a specific response. Red foxes, introduced by Europeans to the Australian continent, have played a disproportionate role in exterminating over twenty species of mammals in the years since European colonization.

Andrea and I built a series of large pens into which we could put a single tammar to study its predator identification abilities. To create an armamentarium of fear-inducing stimuli, I acquired taxidermy mounts of foxes, cats, and wallabies. Because I could not find the skin from the extinct thylacine (a marsupial wolf), I painted a foam model to resemble a thylacine. Then we placed the mounts on a cart and rolled it to the side of the pen, presenting the wallabies with the stimulus.

By placing a small pile of oats at the center of the pen, we baited a tammar to forage. As the tammar was eating, we video-recorded its response to the sudden presentation of the cart, and the wallaby, the cat, the fox, or the thylacine. Despite not having evolved with or having had previous experience with red foxes, tammars responded fearfully to the foxes. Upon seeing a fox, they stomped their hind feet in alarm and skipped to the farthest part of the cage, away from the fox, whereupon they looked at the stuffed fox. Interestingly, the tammars also responded fearfully to the cat, but not to the wallaby. They also didn't respond to the thylacine model, even though it was the physically largest stimulus. So, we found that size alone was not a main determinant that elicited fear.

What could explain the tammar's fear of the fox? With a different tammar population I aimed to find out. Nineteenth-century New Zealand was an island archipelago with very few indigenous mammals. So few that the governor of New Zealand imported them from other places. Among other mammals, he introduced tammar wallabies. This act could be viewed as an environmental tragedy, but the wallabies didn't overrun the island as did many other mammals (and birds) introduced to New Zealand. Moreover, in retrospect, this introduction saved a population of tammars from extinction. How? Because at the same time Governor George Gray was moving the tammars from the mainland South Australian population to New Zealand, the South Australian population was being driven extinct by cat and fox predation. So unlike the Kangaroo Island population, the mainland South Australian tammar population was exposed to mammalian and avian predators before being moved to New Zealand 130 years before. How might an evolutionary history of exposure to predators influence their ability to respond to the sight of predators?

I duplicated our predator recognition experiment in New Zealand on animals descended from this predator-rich South Australian mainland population. Recall, though, after moving from a predatory environment, these individuals subsequently spent 130 years in the suburban, predator-free safety of New Zealand. Remarkably, the New Zealand tammars had seemingly lost their predator recognition abilities, or at least their ability to respond fearfully to any of the models. Why did they no longer respond to the models? What was different? I hypothesized that a predator recognition template had essentially disappeared through disuse. And this is where Andrea's tammar-training work comes into play.

Andrea's goal was to begin with the Kangaroo Island tammars' antipredator response for foxes and enhance it by training the tammars to associate the sight of a fox with something even more unpleasant: Andrea, wearing a witch's hat and mask while chasing the tammars with a capture net! In four paired presentations, the tammars successfully increased the magnitude of their fearful response. They fled in fear every time they saw a fox, because the fox predicted that a costumed and net-waving Andrea was not far away. Thus, Andrea's work demonstrated that tammars, which had a seemingly innate fearful response to foxes, could learn to be even more fearful of foxes . . . and Andrea.

Next, Andrea used a taxidermy mount of a goat that was the same size as the fox. Unlike her experience with the fox, she was unable to successfully train tammars to respond fearfully to the goat despite chasing them with a net while wearing the hat and mask. This result, combined with the results above, strongly suggests that the tammars have a rather specific predator recognition template for a fox.

This is remarkable because, as mentioned above, the Kangaroo Island tammars have had no experience with foxes (until our experiments). Indeed, for the past 9,000 years, the Kangaroo Island population probably had limited exposure to any terrestrial predators. Yet they presumably had evolved from a mainland population that was exposed to mammalian predators, and they had to deal with predatory eagles. What might maintain the persistence of what must have been a mammalian predator-recognition template? A template that without any predators, as our New Zealand results suggest, can disappear in as few as 130 years? But also a template so useful that it enabled tammars who continued to live with eagles to respond to foxes—an entirely novel mammalian predator.

By studying antipredator behavior of tammars, kangaroos, and wallabies on Australian islands, I found that locations where animals had exposure to predators maintained some degree of antipredator behavior, while those populations that had lost all of their predators lost their ability to identify or otherwise respond to predators. One antipredator behavior we focused on with tammars was whether they perceived safety by associating in larger groups of other tammars. These group-size effects were found in a mainland Western Australian population (a population exposed to a variety of terrestrial and aerial predators) as well as on Kangaroo Island (where tammars had to avoid eagles) and on Garden Island (where tammars had to avoid snakes). And in New Zealand (where tammars had no predators for 130 years), they were less wary and did not seem to benefit by associating with others. It makes sense: if any predators are around, maintaining vigilance and perceiving safety in numbers is useful. But what was unexpected is that the presence of *any* predator could maintain the ability to respond to other predators as well. Recall, Kangaroo Island tammars exposed to eagles were able to respond to foxes.

Animals have multiple predators—they have to avoid predation from above, alongside, and below. They have to respond to terrestrial killers, aerial hunters, and sometimes aquatic predators. Some must also avoid getting evicted from the safety of their burrows by predatory bears or badgers.

For species with more than one predator, which is most species, the presence of a single predator could maintain antipredator behavior even for predators that were no longer present. I call this the multipredator hypothesis. How can prey find themselves without some or all predators? The prey could have colonized a location without predators, or some or all of their local predators could have gone extinct. How might the multipredator hypothesis work? If an animal has to avoid predation by different sorts of predators, it would be best to evolve the ability to respond to both potential predators. For example, coyotes stalk on the ground and eagles strike like a lightning bolt from above. If an animal responds well to the threat of coyotes but is terrible at responding to eagles, it will leave no descendants. Thus, animals living with multiple predators simply need the whole package to be, well, whole.

Consider a spotted fawn, a young deer, spending its early days lying quietly, waiting for its mother to come back and feed it. The crypsis from its spots gives the fawn the ability to blend into its background. Combine this with the fawn's immobility, and we have a potent antipredator package of traits. Independently evolving these makes no sense—an immobile but not cryptic animal would be killed, as would a cryptic but mobile animal.

It's been over a decade since I proposed the multipredator hypothesis and most but not all studies have found some support for it. This idea, admittedly stimulated by my studies of tammars, explains how fox-naïve (Kangaroo Island) tammars recognize foxes and why New Zealand tammars do not.

This idea may also explain a lot of human behavior. Why is it so difficult to eliminate our fear of lions, tigers, and bears? According to some analyses, humans have never before lived in such a world so safe from violent death. Not only have we killed many large carnivores that may have taken the occasional person, we've also reduced the rate of death from wars and armed conflicts. Yet, we come from a long lineage of those who were prepared to fight or flee at a moment's notice, and human conflicts have not been completely eliminated. The multipredator hypothesis would predict the persistence of antipredator behavior—in this case our fears for species that we no longer encounter. I suggest that we retain our fear of lions, tigers, and bears because our HPA axis is primed to respond to other threats that we still encounter.

We have also learned from animal studies that looming objects are scary. Imagine an eagle swooping down on a tammar or a rabbit. Wise prey should flee with sufficient time to avoid a fatal impact. Thus, it's essential that animals estimate the amount of time to contact (or impact) of objects. Some researchers have used this foundational response to try to figure out how to prevent animals from hitting motor vehicles and aircraft, which may travel much more quickly than anything animals have ever experienced before. I have capitalized on this escape response to study these responses in kangaroos, birds, and hermit crabs.

Our hermit crab studies began in the Virgin Islands. To set the scene for this study, it is important to note that every other year I teach a very intensive undergraduate field biology course at UCLA. We take our students around the world to conduct novel behavioral research. We begin by assigning students to random groups of three. In their group they must come up with a novel question that I help sculpt into a hypothesis we might be able to test during our three-week field excursion. Many students are concerned about the loss of plants and animals due to our relentless development and anthropogenic assault on nature, so student projects often focus on questions of wildlife conservation relevance. Excessive predation is associated with population extinction, and many projects focus on how animals assess risk or communicate about risk. By studying these topics, my students have contributed to the fundamental literature of how animals avoid getting killed and of their escape strategies. I have chosen the most relevant of these studies to highlight in the coming chapters.

My students—Alvin, Paulina, and Sonja—were interested in asking whether and how human noises distracted animals. This is an excellent question for students to explore because it requires an understanding of attention. Attention, as psychologists tell us, is something that is finite. We can focus our attention on specific tasks, but by doing so, we're unable to focus on other tasks. Our focused attention is also tenuous—we are distracted easily. Such distractions may be costly to us—we can't text and drive well—but are also likely to be costly to non-humans. Animals may fall prey to predators if their attention is unwisely focused elsewhere. My students therefore asked the following question: Do boat motor noises distract terrestrial hermit crabs from paying attention to approaching threats?

Terrestrial hermit crabs live like shipwrecked sailors in the Virgin Islands. They spend their days eating fallen fruits and wandering around the beach, forests, and steep slopes. We planned to see what happened when we interrupted their solitude with sound. We hiked trails and beaches and, upon encountering a hermit crab, either broadcast the sound of motorboats revving through a portable speaker or watched in silence. Then we walked toward the crab until it pulled fearfully into its shell. (I felt bad about this part because if a crab was on a steep slope, it would roll down the slope and then later have to climb back up.) We found that we could get much closer to the crabs when we played the boat motor noise in the background. Other experiments showed that their reaction was not a result of ground vibrations when we walked toward them; when a stuffed shirt loomed silently over them, the crabs were still distracted by the boat noise. Then we conducted a "disco-party boat" experiment, where we paired a strobe light with the boat motor noise and confirmed that more stimuli were even more distracting. What we were really studying was an attentional process in hermit crabs.

At the time, one of the students studying hermit crabs, Alvin Chan, was working in my colleague Aaron Blaisdell's psychology lab studying cognition in pigeons and rats. Alvin wondered if we could have better experimental control over the scenario than was possible in the field and suggested that we bring the question inside. Thus began a wonderfully productive cross-disciplinary collaboration. In this case, bringing the question inside meant starting a small hermit crab colony and then creating looming, computer-generated images to scare the crabs. We initially created an animation of a dot expanding into a large black circle, but found that an animation of an expanding coconut crab scared the crabs into their shells. (Coconut crabs are huge hermit crab predators where they coexist.) We then replicated and expanded on the earlier field results, demonstrating that crabs are indeed distracted by noises—even more so if the noises are loud. This is exactly what we would expect if the crab was distracted. Most of us can commiserate: it's much harder to read a complex passage in a loud public train station than in a quiet home.

My takeaway from these crab studies is that how we respond to fearful stimuli is influenced by how we focus our attention. If we're always looking for scary things, we'll find them. And we'll be scared. Pick up any newspaper and think about the horrible things reported—muggings, rapes, and break-ins. Why do we ever leave the safety of our houses? Yet the incidence of these events is relatively rare. If we focus our attention elsewhere, we will not be so scared. The next time you find yourself scared about something, try distracting yourself by listening to music.

It is also important to note that the crabs hid from the expanding image of a coconut crab because the expanding image predicted the time to contact. Humans predict time to contact as well. When we cross streets, we use our binocular vision to help estimate the speed at which cars and trucks are approaching us. Many animals—deer, geese, and squirrels—need to predict the speed of looming vehicles on the road as well. And birds have vehicular predators on the ground and in the air.

The film *Sully: Miracle on the Hudson*, starring Tom Hanks, recounts the true story of US Airways Flight 1549, which hit a flock of Canada geese. The impact of the geese caused total engine failure and prompted a remarkably smooth response by pilots Chesley Sullenberger and Jeffrey Skiles that resulted in a safe landing. While hitting an entire flock of birds is unusual, individual bird strikes are all too common. Between 1990 and 2013, the US Federal Aviation Administration estimates there were approximately 142,000 bird-airplane strikes in the United States. These accidents led to twenty-five fatalities, 279 injuries, and 639 million dollars of damage for US civilian aircraft. My good friend and colleague Esteban Fernández-Juricic has focused on studying, from an animal's perspective, how animals identify approaching vehicles.

Responding immediately to an approaching object is not always the best strategy for an animal. Often the best first response is to take the time to assess the true risk. For instance, if an animal flees whenever it sees anything rapidly moving toward it, this means it will flee from both attacking predators and nonpredators that are just running or flying by. Such indiscriminate flight means that individuals will allocate precious time to escaping from nonpredators, which is costly in the long run. (We will learn more about the trade-offs animals make assessing risk in Chapter 6.) But when faced with a speeding car, truck, or airplane, animals have limited time to collect important information about risk.

Working with turkey vultures, since they are easily baited with dead deer on the side of the road, Esteban and his colleagues planned experiments to see how the vultures identified approaching vehicles as they foraged. When the researchers drove trucks directly at feeding vultures, they fled at greater distances as the velocity of the approaching truck increased, as would be expected to avoid collision. But they only did so to a point. When the researchers drove faster than about fifty-five miles per hour, the vultures no longer increased the distance they fled as a function of approach velocity, and some vultures were almost struck by the truck. Esteban and his colleagues suspect that the vultures were simply unable to estimate the time to impact for such fast-moving vehicles.

Esteban then brought the vulture experiment into the lab to address a different issue, that of birds colliding with airplanes. For this purpose he and his colleagues studied cowbirds, a flocking bird often implicated in aircraft strikes. They created computer-generated vehicles that approached the cowbirds at speeds ranging from about thirty-five to 223 miles per hour. Cowbirds flew away at greater distances as the simulated vehicle moved faster, but only up to about ninety-three miles per hour. After that, they seemed insensitive to further increases in approach velocity, much like the turkey vultures.

An animal's perceptual abilities have evolved to maximize survival. This explains why vultures and cowbirds have a constraint on their ability to estimate time to impact, as they did not evolve with such fast-moving predators. For these and other birds, threats didn't move at airplane speeds. They haven't had the opportunity to estimate time to impact for such high speeds, nor had they historically benefited from such an ability. Esteban and his collaborators are now exploring ways of making planes and other vehicles even more noticeable to the birds. By combining detailed studies of how specific animals sense and perceive approaching objects and using this sensory knowledge to design specific deterrents, they aim to reduce vehicular collisions.

Fears can be triggered by seemingly small things that in the past were reliable cues of existential threats. This hypersensitivity to small things allows us to identify visual features that are scary: the sight of fear is looming and has eyes, or perhaps it slithers, or has eight legs, or looks like a coconut crab, depending on the evolutionary history of a particular species. Predators are not always visible, of course, and in Chapter 3 we'll explore the role of sound in causing fear. We retain these fears because they have been useful in the past. As long as we have something to fear, they will likely be useful in the future.

3

NOISE MATTERS

It was an early morning flight from LAX to Victoria, British Columbia. I was tired and nervous. My colleagues Liana Zanette and Michael Clinchy met me at the airport, and we drove to a quaint harbor near Sidney, British Columbia. Their inflatable Zodiac bobbed expectantly at the mooring. I'd been dreading this moment since they invited me to visit their study sites on the Gulf Islands because I knew their boat was small, and I knew that the waters off Vancouver Island were cold. A human suddenly immersed in the chilly winter waters does not have much time to get out of the water before getting hypothermia. I didn't sleep well the night before, imagining their small boat capsizing and my body floating hypothermic on the surface before meeting my shivering death.

To my great relief, a practically new immersion suit was waiting for me at the boat. "You're required to wear this for safety reasons," said Michael before he helped me into the bulky red suit and showed me how to seal the wrists and ankles. The immersion suit would keep me visible and dry while delaying the onset of hypothermia if I should find myself in the water. He handed me a life jacket to wear over the suit and then asked me to carry the heavy lead-acid car batteries to the boat. We were going to swap out the batteries required for Michael and Liana's experiments on the islands. With Michael at the helm, the heavily laden Zodiac bounced through the current on our way to Brackman Island, one of five tree-covered islands where they were studying whether the sound of predators could reduce song sparrows' reproductive success.

In this chapter we will learn that prey are able to recognize the sound of their predators, and these sounds alone influence an animal's perceptions of safety and may influence its reproductive success. We will learn that noisy sounds are particularly alarming to a variety of species, including us. This fact can explain why we respond fearfully to certain types of films. We will end with a warning about how easy it is to manipulate the public with fearful sounds and images—something that should inoculate us against the desires of fear-mongering politicians.

Once my colleagues and I reached Brackman Island, after riding through the Haro Strait, Michael drove the Zodiac to the beach. We splashed ashore, offloading the heavy lead-acid battery packs. In addition to motion-triggered video recording systems that ran twenty-four hours per day, seven days per week, there were MP3 players connected to speakers in weatherproof boxes. We carried all of the gear from the boat to their survey site.

The experimental design was elegant. Before the sparrows built their nests, Michael and Liana deployed speakers, positioning them throughout each island, and then began the playbacks. Once a nest was erected they immediately surrounded it with mesh large enough to let the sparrows in, along with battery-powered electric fencing to keep raccoons out. This fencing eliminated any physical effects of predation—hawks, owls, ravens, crows, and raccoons—on egg and chick survival. Sounds of predators would be the only effects.

Michael and Liana created a set of vocalizations from the suite of predators (hawks, owls, ravens, crows, and raccoons) that sparrows could potentially encounter during the day and night. The sounds were broadcast at the appropriate time of day. They also created a set of control sounds, also broadcast at the appropriate time of day, that were similar in terms of the acoustic frequencies and durations but were not threatening. These were frog calls, wind and surf sounds, geese, and loons, as well as hummingbirds and woodpeckers. They broadcast the predator sounds to one set of nests and control sounds to another every few minutes for twenty-four hours per day. To avoid habituating the birds to the playback and thus dulling the birds' response, sounds were played consecutively for four days, followed by four days of silence, throughout the nesting season. Michael and Liana also set up video recording devices to study control nests, those that were not exposed to predator sounds.

Then they waited. After watching thousands of hours of video recordings, Michael and Liana discovered that birds who nested in areas exposed to predator sounds had 40 percent fewer offspring. First, this was because they laid fewer eggs. Then, the eggs laid were less likely to hatch. And, for those eggs that hatched, the young were less likely to survive and leave their nests. This final result, called "fledging success," could be attributed to the parents making fewer feeding visits to their nests: the parents were scared! The birds that heard predator sounds persevered but ultimately were not too successful. They were scared off from both incubating their eggs and feeding their young. And it gets worse. Based on years of work with these sparrows, Liana and Michael knew that food-deprived young were less likely to survive to reproductive age, and if they did, they were likely to have neural and physiological problems affecting their adult survival. Fear-inducing sounds alone had profound effects on the population growth and survival of these sparrows.

Liana and Michael's experimental results corroborate correlative results from other species such as elk and wolves, dugongs and hammerhead sharks, and the very well-studied population cycles of snowshoe hares. All of these studies point to the importance of how perceptions of risk can change behavior and habitat use, affect reproductive success, and ultimately alter population growth rates. Chronic stress has profound influences on humans as well, including on our young. Children reared in stressful, low socioeconomic conditions—particularly those in which they face income disparities—experience increased risks of physical and mental health problems throughout their lives. Growing up poor is associated with emotional and behavioral difficulties that include depression, anxiety, and suicide.

Poverty is linked to higher infant mortality, a higher body mass index, and higher overall mortality rates. Mortality results, in part, from being more likely to have chronic diseases as well as being more likely to suffer from Alzheimer disease later in life. It is clear that chronic stress deeply affects both our and many other species' health and survival.

In Chapter 2 we found that recognition of looming objects is one example of an innate behavior. Prey may be able to identify their predators at birth. But prey also learn to respond to their predators. Associative learning—whereby animals change their behavior based on having survived bad experiences—is a powerful mechanism by which animals modify their behavior to increase their chances of surviving and leaving descendants. Prey may innately respond to the sounds of their predators, or they may learn from them.

To study how animals respond to the sounds of predators, my students Alex Hettena and Nicole Munoz and I studied mule deer in Colorado. Mule deer's predators include coyotes, mountain lions, and wolves. Importantly, while coyotes are common, it is extremely rare to see mountain lions where we work. And wolves, a historically important predator, were extirpated in the early part of the last century at our site in Colorado.

Our study started before dawn. Alex hiked to meadows around the field station, searching for deer. Once she spotted a deer, she slowly and quietly stalked to within about forty meters of the deer and observed its behavior. Since we were most interested in how wary the deer were, we noted each time the deer looked up and around and each time they ran or stotted, a stiff-legged prance that they engage in when they're alarmed. After a short period of quantifying the deer's behavior, Alex broadcast one of several sounds through a small portable speaker. The sounds were from a coyote, a wolf, and a mountain lion. We also broadcast the song from butcher birds, an Australian species not found at our study site, since they are entirely novel to the deer. By including the butcher bird as a control sound, we were able to see if novelty, per se, influenced antipredator vigilance. We found the mule deer were most responsive to wolves and coyotes, but not to mountain lions or butcher birds. From these results we concluded that deer are able to respond to the sounds of at least some of their predators, even extinct ones (the wolves).

To put these results into a larger context, Alex, Nicole, and I reviewed scientific literature on prey responses to the sounds of their predators. We found 183 studies on amphibians, birds, fish, mammals, and reptiles that compared a prey's response to the experimental broadcast of a predator vocalization and a nonpredatory control vocalization. Most of these studies focused on prey responses to predators within the prey's environment, extant predators, and thus species with which they had potential prior experience. Recall, we included wolf vocalizations in our mule deer study because we wished to learn whether the deer could respond to an extinct predator. Our review showed that while most species responded to the familiar predatory sounds, it was less common for prey to respond to the sounds of extinct predators. Therefore, experience living with predators plays some role in knowing what to fear.

But what is it that permits animals to respond to the sound of predators with which they have no lifetime experience? To study this, my colleagues and I conducted a series of experiments on a species I've studied for over twenty years, yellow-bellied marmots. As in earlier experiments, we asked whether marmots could respond to vocalizations from extinct predators (wolves) and extant predators (coyotes and golden eagles). Previous experiments revealed that marmots don't respond to novel sounds, so we used an alarm call—a vocalization animals (including marmots) emit when they detect a predator—to see just how scary predator sounds were.

To do this research, we baited marmots to a location about one meter from the safety of their burrow and ten to twelve meters from a hidden speaker with a handful of horse feed. Sitting quietly, often for hours, we waited until a marmot began to feed on the bait. Since we had marked all of our marmots with dye on their backs, we could tell individuals apart. Once an animal began foraging, we recorded its behavior, and then, after a time, we broadcast the vocalizations from the hidden speaker.

Individual marmots responded most frequently and intensively to the alarm calls: they looked around and suppressed their foraging. Interestingly, vocalizations from wolves and eagles but not coyotes also elicited responses. To see if there was something about the nature of the vocalization itself, or whether marmots had evolved some sort of recognition template, we changed our experiment. We broadcast the eagle calls backward and compared the response to that evoked by a normal forward call, since the exact same frequencies were present in the forward and backward eagle call, but the tempo and frequency trace is different.

As previously discussed with respect to detecting visual stimuli, recognition templates are expected when animals must make rapid decisions. But unlike the eye spot templates, which would have predicted that certain frequencies or other characteristics of the sound might be particularly evocative, we found that marmots needed to hear the calls in a certain way—in this case, forward. The result of a reaction to forward eagle vocalization, with its particular tempo and frequency, showed that recognition templates can be very specific.

We wanted to better understand how the marmots responded to their extinct predator, wolves. Was it something about the lower frequency of wolf howls, or was it something about the length of wolf howls that elicited a response? Wolf howls are naturally of longer duration and lower frequency than coyote howls, which are "yippier" and higher pitched. Could it be that marmots are particularly sensitive to howl duration? As it turns out, the answer is yes. When we compared marmots' responses to normal-length wolf howls and exceptionally long coyote howls, which were of similar duration to wolf howls, we found no difference in marmots' responses, despite the fact that the wolf howls were at a lower pitch. Thus, by focusing on howl length, which is typically longer in wolves than coyotes, marmots are able to respond to wolves.

In another experiment conducted in Australia, we found that mobbing vocalizations, calls seeking help to chase away predators, can elicit responses in multiple species. Mobbing calls are associated with the mobbing behavior we discussed in Chapter 2. The vocalizations are broadband, rapidly paced, pulsatile tones, and they sound similar, even across species. If you see a crow surrounded by songbirds and hear a cacophony, the songbirds are likely giving mobbing calls. Through an undergraduate field biology research project, we found that pulses of sounds containing the dominant frequency of these broadband mobbing calls are as effective as real mobbing calls at eliciting attention in apostle birds, a highly social Australian species. Pulses of broadband white noise did not elicit attention. This finding focuses our search for the specific structure of templates.

When results of the deer, marmot, and apostle bird experiments are considered together, we see that animals can have specific acoustic predator recognition templates, much as we've seen before with visual predator recognition templates. The specific frequencies heard, the tempo of a vocalization, and the vocalization's duration may all enable animals to respond to their predators. But how specific are these responses and how specific is the information that animals acquire from the sound of their predators?

One year, two groups of our students studied risk assessment in dik-diks—a small, gracile antelope—in Kenya. In East Africa, dik-diks are eaten by about thirty-six different species of birds and mammals. Dik-diks should, and do, live in fear. They are monogamous and spend their days next to their spouse, often in the open, looking, listening, and smelling for any sign that a predator is nearby. Their response, upon detecting a predator, is remarkably sophisticated.

One of our student groups broadcast the sounds of eagles to dik-diks. Upon hearing these sounds, the dik-diks looked up and scanned the sky for risks from above. The other student group broadcast the sounds of jackals. Upon hearing these sounds, dik-diks looked around and wiggled their somewhat prehensile noses—sniffing for smelly mammalian predators. Similarly, several species of forest-dwelling monkeys from West Africa have shown remarkable abilities to respond appropriately upon hearing leopards, predatory baboons, or monkey-eating eagles. Just as seen in visual predator recognition templates, dik-diks and the forest-dwelling monkeys evolved the ability to recognize sounds from their predators, whether from the air or ground.

But what is it acoustically that makes sounds scary? I gained new insights into the acoustics of fear-inducing sounds one day while trapping marmot pups. Over years of experiments I've studied eight of the fifteen species of marmots around the northern hemisphere, but it was an experience at the Rocky Mountain Biological Laboratory that refocused my research program.

To undertake our research on marmots, my co-investigators and I follow marked individuals throughout their lives to understand how animals cope with a variable environment, both individually and as a population. Animals have personalities, much as humans do, so we ask questions about the value of those individual differences within a species. We ask questions about the value of social relationships and how social dispositions vary over their lives. And we can ask whether traits like these are heritable. We study factors that explain differences in the survival, reproductive success, and longevity of animals, and we ask detailed questions about their antipredator behavior. To mark animals we bait walk-in wire mesh traps with horse feed. Once trapped, the marmots are quickly weighed, given ear tags if required, and given distinctive marks on its back for identification purposes using an old toothbrush and dye. We check each marmot's reproductive status, collect feces if forthcoming, and, if it is an older animal, collect a small blood sample. We aim to trap marmot pups as soon as they've emerged from their natal burrow because they quickly become favored prey to foxes, coyotes, and raptors. They are so favored that only 50 percent of them survive to the next year.

One day while being gently held after capture, a marmot pup looked at me, opened its mouth, and screamed. I was so startled that I almost dropped it. But I'd learned the hard way never to let go of an animal, even if it is biting you, because every capture could be the last for that individual. I'd never had such an experience with a marmot pup before, and I'd never heard such a horrible sound. What was it about the sound that was so emotionally evocative? I'd never had this response before to a marmot alarm call, even if the calls were from pups.

In response to my visceral reaction my colleagues and I began to study pup screams. We noticed that sometimes following a pup scream the pup's mother emerged from their burrow and approached us. A pup's mother did not usually have this response when pups alarm called; typically animals flee to their burrow and remain there in safety when there is an alarm vocalization. We quickly learned that screams differed from alarm calls in pretty much every acoustic dimension: screams are longer, have a lower frequency, and are characterized by abrupt frequency jumps; screams sound noisier than alarm calls. It was the noise that caught my attention.

A pup scream reminded me of what a car stereo sounds like when it is turned up. The music on the stereo first sounds louder, and the fidelity remains high. But at some point, if you keep turning the volume up, the music begins to sound distorted—noisy—and there may be rapid frequency fluctuations. This distortion and these rapid frequency changes also are produced when you overblow a trumpet or when a marmot overblows its vocal chords. To understand why, we have to take a brief divergence into thinking about nonlinear dynamic systems.

When nonlinear systems transition to nonlinear outputs, such as when the music coming out of your speaker begins to sound distorted, they behave in predictably unpredictable, chaotic ways. Chaos is a branch of mathematics that is readily applied to all sorts of dynamic systems—those that vary over time and have a variety of inputs or factors that may influence their outputs. Chaotic systems are especially sensitive to slight changes in the value of their inputs. Vocal production systems are complex dynamic systems. Indeed, nonlinear vocal production systems produce deterministic chaos, one type of nonlinearity that emerges when inputs change slightly. Thus at a lower volume your speaker sounds okay, then suddenly, as you turn it up just a little more, it shifts and doesn't sound so great. Nonlinear vocal production systems also produce the rapid amplitude fluctuations (the volume sounds like it's going up and down) and frequency fluctuations (the tone changes rapidly) that characterize a blown-out system.

Animal and human screams are chock full of these nonlinear acoustic attributes. The first scream that the actress Janet Leigh emits in the classic shower scene in the film *Psycho* is a classic example of noisy and nonlinear sound. By contrast, when Marlon Brando screams "Stella!" in *A Streetcar Named Desire,* it's not a fearful scream. It's not noisy. I wondered why marmot screams were full of nonlinearities but alarm calls were not. I became curious to learn more about nonlinear sounds across the tree of life. What is the function of nonlinearities in animal vocalizations?

A former student and now colleague, Zach Laubach, shared commercial recordings of prey screams. These recordings are used by hunters aiming to trap or kill predators because predators are attracted to the screams of their prey—rabbits, foxes, and deer in distress. I found the tapes difficult to listen to because I suspected the sounds had been elicited by torturing the animals. These recordings were full of nonlinearities. Further, I found recordings of chimpanzees giving their classic pant-hoot display, overblowing the crescendo and creating noisy, nonlinear sounds. The meerkat, a southern African mongoose, produces progressively noisier alarm calls as it becomes more scared. Piglet squeals and rhesus macaque screams have similar characteristics. Through this study I realized that there is a theme: all of these species, when in similar aroused and fearful situations, emit nonlinear, noisy sounds. Upset dog barks are raspier, noisier. It led me to ask, is noise the sound of fear?

To explore this question, we conducted a series of experiments on marmots and birds. We inserted a little bit of noise in normal marmot alarm calls and as a control removed the same duration of sound to create a silent gap. Both calls were nonlinear, with rapid amplitude and frequency fluctuations. How did the marmots respond to the calls with a little noise added? When hearing calls with noise added they suppressed foraging and spent more time looking around.

We studied birds' reaction to added noise in the mostly uninhabited Calabash Caye in Belize. My undergraduate field biology students and I broadcast three different kinds of tones: pure tones, tones that immediately shifted up in frequency, tones that immediately shifted down in frequency (like hitting two piano keys in rapid succession), as well as a pure tone with a bit of noise added at the end. Caribbean great-tailed grackles, a common bird on the small island, did not respond to the pure tone or to a control bird call, yet responded the most to the noisy playback. Back in Colorado, a student and I repeated this experiment on white-crowned sparrows, a common bird around the field station. White-crowned sparrows had the same response as the great-tailed grackles. Noisy sounds, even completely artificial ones, appear to be evocative. Recent results from another field biology group studying a nonvocal skink on the island of Moʻorea in French Polynesia have shown that even nonvocal animals have the ability to selectively respond fearfully to synthesized nonlinearities and noise.

I discussed some of these research results during a public talk at UCLA and speculated that the noisy sounds may be evocative to humans as well, noting that perhaps musicians and film score composers capitalize on this to make scenes more disquieting. To my delight, Peter Kaye, a musician and film score composer, introduced himself to me during a break. Peter was developing a biological basis of how music affects our emotions. He introduced me to a huge music theory literature that is essentially devoid of a biological basis for our emotional responses to music. Peter found my hypothesis interesting, and I suggested a collaboration. We recruited a UCLA honors student, Richard Davidian, to work with us. Using Internet "best of" lists, we developed a list of the best horror films, the best sad dramatic films, the best action / adventure films, and the best war films. Then we found iconographic scenes in these films, such as the shower scene in *Psycho*, or the scene where John Coffey, the wrongfully convicted and innocent African American man executed in *The Green Mile*, walks to the execution chamber. We extracted a thirty-second acoustic clip of each scene and made spectrograms—voice prints—of these clips, then analyzed them.

Like dynamic systems, music soundtracks are complex. They may contain dialog, song, Foley (sound) effects, diegetic (natural) sounds, as well as music from one or more instruments. At first listen, there was nothing in the soundtrack that reflected a system—blown out or not—due to the hours of music production. Could we find *analogs* of nonlinearities? Could we find instances where there was noise or rapid amplitude or frequency changes?

Peter, Richard, and I spent a lot of time scrutinizing spectrograms and developing standardized ways to score them. Once we were happy that we could do this consistently, Richard went to work listening to the clips while looking at them. In each clip he scored the presence or absence of specific features, including (but not limited to) noisy screams (we differentiated between those produced by males and females), noisy sound effects, and abrupt frequency changes.

We found that sad films suppress noisy sound effects and enhance abrupt musical frequency changes (cue those dynamic violins, which are quite effective at getting us to cry). In comparison, horror films suppress abrupt musical frequency changes and are uniquely characterized by noisy female screams. We had found correlative evidence that noise is used to influence our emotional responses in films.

I discussed these results with my UCLA colleague Greg Bryant. Greg's field of study is emotional communication, and he is also an accomplished musician. Intrigued by our correlative findings, I asked Greg if he could help us conduct experiments on humans. Let me let you in on a secret: much of what we know about human psychology comes from experiments conducted on willing undergraduate university students who get some course credit for participation. With student volunteers signed up, we prepared for the experiments.

To begin we filmed ten-second video clips of people doing rather benign things, like walking or picking up a telephone. For instance, a woman walked down the road, and after five seconds turned to cross a street. In another video clip a man was sitting in a chair, and at the five-second mark picked up a ringing telephone. In yet another clip, a woman sat in a chair at a table, and after five seconds took a sip of coffee. Peter and Greg composed ten-second marmot-inspired music, a bit like Muzak. However, at the five-second mark either the music continued or we added noise or another type of nonlinearity to the musical composition. We added our Muzak score to the video clips. Now we were ready for the student subjects.

Student volunteers were played music by itself, the video clip alone, or the video clip with musical score. Each subject heard many different compositions and watched many different clips. We asked the volunteers to describe how positive what they saw and heard was on a scale that went from negative to positive numbers, and how aroused they felt after being presented with each experimental stimulus. Noisy music, when played without the video clips, was the least positive and most arousing, as perceived by the students. Using what I like to refer to as marmot-inspired music, we could evoke emotional responses in a way that supported the nonlinearity and fear hypothesis: the sound of fear is noisy!

Interestingly, the addition of the video clips seemed to attenuate the responses to the music. The volunteers no longer assessed the videos with noisy music as emotionally evocative. While unexpected, in retrospect this makes some sense. When presented with an acoustic and a visual stimulus, the visual stimulus became dominant. Readers old enough to remember film reels are well aware of the situation where the soundtrack and image get separated. As viewers we are able to compensate for a while by lip reading until we are unable to perceptually bind the different stimuli anymore. What we had created in the experiment was a mismatched, multimodal stimulus. The visual channel was benign (by design), and the acoustic channel could have been alarming (again by design). In this case our students weren't visually deceived, thus they didn't respond to the benign videos coupled with potentially scary sounds. But when the sounds were played alone, the music was particularly evocative.

We continued these studies by repeating the same experiment, but with one addition. The second time we recorded electrical activity in the students' facial muscles. There's a large literature going back to Darwin on the universality of emotional responses in humans and nonhumans. Darwin described fearful animals that produce high-pitched vocalizations as shrinking back—imagine a scared cat arching its back or a scared puppy trying to make itself look small while backing away. By contrast, aggressive vocalizations and postures are the opposite: loud, noisy, and associated with an inflation of posture, such as seen when an aggressive dog or gorilla puffs up its fur and charges its opponent. Darwin described a similar response for human facial expressions. Later work quantified the muscles that fire as people experience different emotions. Our experiment focused on whether these muscles are fired when triggered by sounds or visual stimuli, as a large body of work suggests. For instance, when you're fearful or surprised, your eyes quickly open wide, as you might already know by watching your friends' responses to scary films. Indeed, our experimental results showed that noisy sounds lead to increased activity of these corrugator muscles and rapidly opened eyes. Thus, our results showed that noisy music leads to physiologically detectable emotional responses.

As discussed in Chapter 1, neuroscientists use functional magnetic resonance imaging (fMRI) to look at neural responses to a variety of stimuli. If shown a scary visual, parts of the brain associated with fight or flight activate. Therefore, the next logical step in our research was to see if our marmot-inspired music activates the part of the brain, the amygdala, that is directly associated with responding to fearful situations.

While we've not been able to conduct these experiments yet, another research group published a fMRI study that showed how scary sounds are rough. Roughness is an acoustic characteristic defined by very rapid, sawtooth-like changes in amplitude. It's important to realize that roughness is not noise, but noisy sounds can be rough. Screams, according to their work, are also characterized by being rough. When the researchers put subjects into an fMRI machine, they found that rough sounds uniquely activated the amygdala. Thus, it appears that the sound of fear is noisy and rough.

So, how can we use these fundamental insights gained by studying marmots and humans? One way is to use this knowledge in creative pursuits, including music. I've had the opportunity to meet some excellent musicians in Los Angeles, and I always ask them to tell me how they create emotional scenes in their music. Classically trained musicians immediately jump to music theory—minor scales are disconcerting, and so forth. While punk musicians intuitively use a lot of noise to capture the attention of their audience, those I've asked have not specifically told me why. Not one musician I've met has stated a biological basis for why we respond the way we do. Armed with a more biologically inspired understanding of the sound of fear, musicians could add specific nonlinearities, such as noise, that will tap into this basic emotion and effectively create music and other sounds that scare us.

But what about use of these insights more broadly? Since the 1950s political TV advertisements have sought to create both positive and negative emotions in viewers, often to scare us into voting for a particular candidate. As we saw in our experiments, noisy sounds and fuzzy images (visual noise) are particularly good at evoking fear. President Lyndon B. Johnson's 1964 *Daisy* advertisement is a perfect example: a little girl counts up in an endearingly childlike way as she plucks petals off a daisy, then a Mission Control–style voice counts down. Suddenly, the camera zooms in on her eye, and the frame is filled by a giant mushroom cloud. Johnson amplifies the dire message in a scratchy voice laid over the video clip. The ad incorporates the techniques discussed here—noisy sounds and noisy images—to stoke public fear of Johnson's political opponent, Barry Goldwater, whose support for the aggressive development and use of nuclear weapons was known.

Certain sights and sounds have been shown to elicit fear in animals and also in humans. What role do our other senses play? I'll explore scents that may incite fear in Chapter 4.

4

SMELLS RISKY TO ME

Michael Parsons worked with chemists to develop artificial dingo urine that would ideally be a mix of biologically salient chemicals found in real dingo urine. Why dingoes? Dingo urine (but intriguingly not dog urine) can scare kangaroos away from preferred foods, such as muesli or highly palatable alfalfa pellets. While this finding may not sound like a game-changer to you, it was a potentially important discovery for Australian mining. Environmental laws in Australia require mines to reclaim their land by replanting vegetation removed during excavation. Kangaroos, it turns out, feast on young plantings and had severely impeded revegetation of areas that were being reclaimed. Thus, effective kangaroo repellents are desperately needed. If dingo urine could be used as a repellent in these reclaimed areas, then it would offer a natural solution to the hungry kangaroo problem. Michael's experiment took place when some Australian states were about to ban lethal control of "problem" kangaroos, so a welfare-friendly kangaroo repellent was a top priority.

To see the effects of dingo urine firsthand, we went south from Perth through the jarrah forests to Roo Gully Wildlife Sanctuary, near the small, rural town of Boyup Brook. In the main house at the sanctuary Michael uncapped the jar of chilled urine, carried to Boyup Brook in a beer cooler, and poured a measured amount into several four-inch petri dishes—the squat, wide glassware you may recall from high-school biology lab. Then he gently mixed in water-storing crystals—typically combined with soil to keep houseplants' roots moist—to create thin, gelatinous hockey pucks of dingo urine. As a control, in other petri dishes he mixed distilled water with the crystals.

The well-habituated kangaroos hopped to the back porch of the main house for the afternoon feeding. In food trays that contained kangaroo pellets and fresh vegetables, Michael put either a dingo puck or water puck. By dusk the western grey kangaroos were feeding only at the bins containing the water pucks. Hours later I went to check the bins. In the bins with the water pucks the food had been consumed, but in the bins with dingo pucks the food was untouched. I'd seen the videos of kangaroos hopping back in fear when they detected dingo urine, but to observe the complete avoidance of these bins was striking. It showed that a predators' scent alone could prevent animals from foraging on a preferred food.

Animals respond fearfully to a variety of olfactory cues associated with predators. But are they specific chemicals or mixes of chemicals that universally invoke fear? In this chapter we will learn about the various chemicals that have been studied, the ways that nonhumans (mostly mammals and fishes) respond to predatory cues, the potential information contained in them, and how they can be used by humans to attract and repel animals.

In a follow-up experiment we established a set of feeding trays at Caversham Wildlife Park, near Perth, Western Australia. As in the previous experiment, Michael put out dingo urine or dingo feces near some food trays and water near others. What was remarkable about this experiment was that the three species we studied there—a mob of red kangaroos, another mob of western grey kangaroos, and two agile wallabies—quickly learned to avoid the feeding trays with urine or feces near them. Indeed, when they approached the predator-scented feeding trays, they fled and stomped their hind feet in alarm. And, somewhat remarkably, after only ten days of presenting these scents, they completely avoided the predator-scented areas. It's important to note that the animals in the wildlife park could safely avoid these areas because they had other places to get food, but leaving such high-quality food behind was remarkable nonetheless.

Michael and I conducted more studies. We sought to understand whether the response to dingo urine was in some way innate; if so, then we would expect that it could work even with dingo-naïve animals. While mainland Australian animals have lived alongside dingoes between three and five thousand years, Tasmania was isolated from mainland Australia by rising sea levels long before dingoes moved into Australia. Thus, dingo scent should be completely novel for Tasmanian animals. Michael and I wanted to know whether species on Tasmania could be repelled by dingo urine. Determining this required some engineering and construction.

We surrounded an automatic feeder with wire fencing. The feeder could only be accessed by an animal pushing a door open. After giving wild Tasmanian pademelons—a small, squat forest kangaroo relative—and brazen, brush-tailed possums six weeks to learn to access the feeders, the experiment began. Dingo urine was placed in two locations to mimic varying levels of danger. It was placed outside the doors in an open space that provided access to the feeder, as well as inside the sheltered feeder very close to the food. Pademelon and possum activity at the feeders was recorded inside and outside the fences on surveillance video over several months. As the nights progressed, members of both species made increasingly more wary and tentative approaches to the feeders. Both species were more likely to flee before accessing the feeder on days when dingo urine was present than on control days when no dingo urine was present. What is it about dingo urine that even dingo-naïve animals find scary?

Well, we're not sure. One of our collaborators, Ken Dodds, works at the Western Australia Chemistry Center in Perth. Ken showed us the various gas chromatographs and mass spectrometers he used to isolate specific chemicals from samples. These machines heat a substance and as specific chemicals in that substance volatilize (become a gas), the released gases are sucked away by a vacuum. The specific temperature at which they volatilize is the first step toward identifying the chemical. The gases are then pumped through a mass spectrometer, an instrument that measures the specific mass of each chemical. The output from this process consists of two lines that, in real time, plot the abundance of chemicals and their specific mass. The first line is flat when there are no chemicals being volatilized, increasing suddenly when a chemical volatilizes, then crashing back down when no more chemical is present. The height of these peaks represents the amount of chemical present. The second line reports the mass of each volatilized chemical. A database with identified chemical profiles is used to try to match the specific chemicals that are contained in the sample. Such information can in theory be used to manufacture an inexpensive synthetic equivalent that would be readily available.

Using this equipment, Ken and Michael showed that dog urine is chemically different from dingo urine. Dingo urine is much more complex; the mass spectrophotometric profile of dingo urine has many more peaks and many more specific chemicals. Dingo urine seemed to also vary by sex, a finding that was not unexpected because urine contains a mix of metabolized chemicals, including sex hormones (testosterone and estrogen). Many studies of intraspecific communication in other species of mammals have shown that animals are able to discriminate among sexes after taking a whiff of urine. It turns out that urine contains all sorts of interesting information about the identity of the signaler.

But urine is not the only detectable scent. Predators, like all animals and humans, have scents on their skin, feathers, or fur, and in their feces as well. Glands, scattered around the body, produce these scents to communicate to other members of their own species. But vulnerable animals become attuned to whatever cues they can acquire about the presence of a predator. From a curious prey's perspective, detecting a urinary scent or fecal scent provides somewhat ambiguous information. The predator could be right there or have passed by hours ago. By contrast, detecting odors from fur, feathers, or skin provides information that the predator is nearby!

We have descended from a long lineage of prey with sophisticated noses. Our ancestors who detected the difference between the bouquets produced by these different secretions were likely to have survived to produce offspring. In fact, whether we are aware of our abilities or not, humans can seemingly identify predators by scent.

I tested this hypothesis with groups of adults and children at the Rocky Mountain Biological Laboratory. After research walks and talks about our group's research on marmots, deer, and birds, I opened up the urine samples that we have stored on shelves. Without divulging which scent is which, I asked individuals to tell me whether a given scent comes from a carnivorous predator or from an herbivorous prey. Almost every person identifies the scents properly: predator pee is more pungent than the urine from plant eaters. Why do predator urine and prey urine smell remarkably different? Mammals that eat meat produce similar chemical waste products from the digested meat.

My students and I have studied olfactory predator detection in marmots and deer at our Colorado field site. To do this we order urine from online suppliers. Hunters and trappers, as well as homeowners desperately trying to protect their flowers and ornamental shrubs from hungry herbivores, drive this market of urine collected from captive coyotes, foxes, wolves, mountain lions, deer, and elk. Hunters and trappers use the urine from prey to attract predators and the urine from carnivores to camouflage their own scents while hunting. We used the urine to test the response in marmots as they foraged, much as Michael had tested western grey kangaroos.

To study this hypothesis in yellow-bellied marmots we put out a handful of horse feed, drove a large nail into the center of the pile of food, and affixed a cotton ball scented with urine from either a predator or a nonpredator. We recorded the rate at which marmots sniffed at the food, the rate at which marmots looked around while foraging, and the rate at which they foraged. The marmots did not respond fearfully to water or to urine from elk or moose, but they did respond fearfully to predator odors, including foxes, coyotes, lions, and wolves. When they caught the scent of a predator, they foraged less and looked around more often. Interestingly, they had the most significant reaction to urine from coyotes, a major adult marmot predator, and from mountain lions, a rare predator most marmots probably had limited exposure to. They also responded to urine from wolves, a locally extinct predator, and from red foxes, a predator that usually ignores adult marmots in favor of recently emerged pups. Thus, marmots seemingly have the ability to assess predation risk by smelling urine. And, because they still respond to the scent and sound of wolves, as discussed in Chapter 3, marmots seem to have retained their fear of an extinct species.

How do we best understand animal and human responses to predatory scents? As a behavioral ecologist who seeks to understand differences in behavior, I view the challenge to understand the variation in responses to olfactory predator cues as a full employment scheme—there's a lot of work to be done. Context is everything, and appropriately responding to a predator will vary based on both internal and external stimulation. In other words, animals and humans are constantly making trade-offs. For instance, you would likely have a very different response to a puppy eating your shoe after watching a comedy film than after arriving at home late due to a traffic jam. Our emotional responses are influenced by our state, and some of the behavioral variation we see in response to scents could be state dependent.

Animals respond to odors when chemicals bind to specific olfactory receptors, much like the way a car key is inserted into a car's ignition. If the key fits, the car starts. Similarly, if there is an appropriate olfactory receptor, the chemical can modify behavior. Are there specific chemicals that work to enhance fear? Is there a scent for danger?

One such promising chemical is trimethylthiazoline, or TMT. This chemical is isolated from fox urine and feces. Captive rats exposed to TMT activate their amygdala—suggesting that it taps directly into the fear system. Much of this olfactory work is conducted by neuroscientists who seek binary responses. They want to know, for example, whether or not a chemical elicits a fearful response from an animal, such as freezing in fear. But TMT and other chemicals being studied elicit a complex and variable set of behavioral responses. Perhaps we should expect greater variation because olfactory signals are more like perfume, with its many different constituent chemicals, than eye spots. For instance, days-old urine is chemically quite different from fresh urine and therefore the message is different—"a predator was once there" versus "a predator is here now."

Another likely chemical that elicits fear is called 2-phenylethylamine, or PEA. (Yes, a chemical found in pee is called PEA.) This biogenic amine is produced in the urine of carnivores but not herbivores. Indeed, carnivore urine contains as much as three thousand times more PEA than herbivore urine. Based on our earlier discussion, you might guess PEA is a product of digested meat, but this has not yet been convincingly determined. We do know that mice and rats have specific olfactory receptors for PEA, similar to the aforementioned ignition system in a car; PEA binds to the receptor like a car key inserted in the ignition, sparking a reaction.

Mice and rats also have a set of trace amine-associated receptors (TAARs), including TAAR4. Just like the olfactory receptor, TAAR4 is activated when exposed to PEA. To study this interaction, neuroscientists created cell lines in the lab that either had or did not have a TAAR4 receptor. Laboratory cell lines with TAAR4 receptors were sensitive to bobcat and mountain lion urine, but not to mouse, human, or rat urine. Do species vary in the number of their TAAR receptors?

Yes! And more importantly, there is variation in the number of TAAR4 receptors in different species: rats have seventeen; mice, fifteen; and humans, only six. This means that some species may be better at detecting and responding to predatory scents. We would predict that rats and mice would do better than humans. Behavioral studies have indeed shown that laboratory-reared rats and mice that were never previously exposed to PEA or carnivore urine avoided areas with either present. Additional studies have shown that disabling the TAAR4 gene specifically eliminates the ability of mice to respond to PEA. Thus, PEA may be the scent of danger for rodents.

It is fascinating that TAAR4 receptors are present in the nose but not in the vomeronasal organ, a patch of sensory cells within the main nasal chamber of amphibians, reptiles, and at least some mammals. Scents like the one in PEA that activates the TAAR4 gene are possibly unlike scents that function through stimulating the vomeronasal organ. The vomeronasal organ contains many chemical receptors but is not connected to olfactory nerves. Thus, we cannot say that animals that respond to chemicals detected by the vomeronasal organ are aware that they "smell" them, but nevertheless chemicals acting there may trigger behavioral changes, as the vomeronasal organ is involved in pheromone detection. When a dog urinates on the base of a tree to advertise it has been there, the chemical message includes information about the dog's sex, identity, and reproductive state. Mouse urine contains all of this information as well as pheromones called MUPs (short for major urinary proteins) that can attract members of the opposite sex. A similar chemical may be present in humans. While humans don't seem to have vomeronasal organs, the jury is still out on whether we have pheromone-based communication or assessment.

A recent discovery suggests that there is an affinity between predators and their prey in their response to a specific chemical found in blood. Specifically, *trans*-4, 5-epoxy-(E)-2-decenal, or E2D, is as attractive as whole blood to blood-sucking stable flies and wolves—both of which are known to be attracted to blood. Mice are as repelled by E2D as they are to whole blood. Mice placed into a box with two chambers spent less time in experimental chambers with E2D or whole blood compared to chambers with various control scents. Humans exposed to E2D have an increased galvanic skin response and move around more compared to humans exposed to control scents. E2D, it seems, is a blood-borne cue that is used by predators and prey alike; it provides information about both food and danger.

We have made a good deal of progress on understanding scents associated with danger through studies of fish. Unlike in air, in water chemicals associated with scents are concentrated and more easily assessed. High concentrations of chemicals in water means that danger is nearby; lower concentrations imply that a threat is far away. Chemical information about risk found in water can flow around objects, unlike visual cues. Fish are aware if something scary is lurking behind a rock. Prey fish can even respond to disturbance or alarm cues produced by other species. It's thought that the ammonia excreted when an animal moves quickly away from a predatory threat warns other prey. But this is nonspecific information. Fish can also identify their predators directly.

Prey fish respond to chemicals naturally produced by their predators, including chemicals in the predator's skin or waste-excreted chemicals. Prey fish respond fearfully when they detect broken fish skin. Why? Because when a fish is attacked or killed the skin is broken, and this process creates chemical alarm cues. Studies have shown that prey fish have a remarkable ability to rapidly learn to associate a predator's natural scent with an alarm cue—such as those found in prey's skin. What and how fish learn to fear will be discussed in detail in Chapter 7. For now, let's just say that when the scent of death is fresh, prey should be particularly wary.

Remember, however, prey are always seeking new ways to avoid predation, and predators are always seeking new ways to be more effective. Eliminating olfactory cues may be one predatory strategy to increase hunting success. For example, some aquatic predators, such as sculpin, appear to be able to mask or otherwise digest key components of the scent of their food to prevent others from detecting excreted chemicals of danger. This olfactory cloaking is another vital strategy employed to increase hunting success.

Some prey have also learned to mask chemicals from their predators. Harlequin filefish eat coral and retain the smell of coral, effectively camouflaging themselves from predatory fish. And gulf toadfish excrete urea rather than ammonia, since the latter attracts predatory fish. Thus, by excreting urea, the toadfish are chemically invisible to their predators.

But what if prey are unable to respond to the scent of their predators? The parasite *Toxoplasma gondii* is capable of infecting all warm-blooded animals. This parasite begins its life cycle by living and reproducing in cats. Cats excrete parasite eggs in their feces that develop and can be ingested by rats. Since the parasite can only reproduce once inside cats, however, the parasite only survives when cats eat the infected rats. But the infected rats are more likely to be killed by cats because toxo-infected rats are unable to detect the smell of cats. Researchers discovered that healthy rats flee when exposed to cat fur, but toxo-infected rats exhibit none of this fear. Humans become infected by toxoplasmosis from their pet cats or from eating undercooked meat. Toxo-infected humans are more likely to die due to trauma than non-toxo infected humans. Car accidents, action sports injuries, and other deaths associated with enhanced risk-taking are more likely among people who test positive for toxoplasmosis. The parasite is able to take over its victims' nervous systems and reduce fear, an outcome that impacts behavior and ultimately survival.

Risk is everywhere, and it's not just tasty herbivores that live in fear in our dog-eat-cat world. Predators are killed by other predators. Thus, predators should be sensitive to the scents of their predators as well. Indeed, intra-guild predation, in which predators kill other species of predators, has a potent effect on ecological communities. The remarkably successful reintroduction of wolves to the Greater Yellowstone Ecosystem has taught us many things, but one is that coyotes would have been opposed to the reintroduction. This is because wolves kill potentially competitive coyotes. As wolf populations have increased, coyote populations have concomitantly decreased. These effects are not restricted to wolves and the smaller-sized coyotes.

Dingoes in Australia also kill smaller carnivores. In this case they're doing an ecological service because the smaller carnivores that they kill are mostly feral cats and European red foxes. Both species were introduced to Australia and have had a disproportionate effect on Australia's endemic biodiversity. Both cats and foxes are responsible for the decimation of native mammals—over twenty species of Australian native mammals have gone extinct since European colonization, and much of the blame is attributed to these two species. But dingoes also eat sheep, and this has led to them being shot, poisoned, and fenced out of much of Australia.

In the nineteenth century, Australians built a nearly 3,500 mile fence to keep dingoes in the arid zone and away from prime grazing areas in southern Australia. South of the dog fence (as it's known), very few dingoes survive. North of the fence, dingoes are doing much better, despite ongoing persecution from ranchers. Here, the dingoes inspire fear in kangaroos, which they eat, and in cats and foxes, which they simply kill. Because dingoes slay the cats and foxes they encounter, more native animals are present above the dingo fence than below the dingo fence.

These findings led me and my colleagues to propose a novel idea. Small predators live in fear of encountering larger predators, so small predators should respond to the scents of larger predators and avoid those areas. If there are fewer small predators in certain areas, then perhaps those places are safer for prey.

Our experiences at the Rocky Mountain Biological Laboratory town site support this hypothesis. We had very few successful marmot litters around the Rocky Mountain Biological Laboratory town site for over a decade because of the foxes. Both coyotes and foxes eat marmots, but the foxes prefer marmot pups. Outside the town site, marmot pups have a greater chance of survival because coyotes prey on foxes. If our hunch is correct, then wise prey are descendants of successful ancestors and should be able to estimate the relative safety of an area by detecting scents associated with their own predators and their predator's predators. And prey should feel safer in more intact ecological communities where their predators, or competitors, have predators.

Some studies have suggested that such complex risk assessments are possible. For instance, wild-caught stoats, or short-tailed weasels, respond differently to body odors from members of the same species than from predators. Researchers quantified stoat foraging behavior in the presence of experimentally provided odors. Stoats were most cautious when they encountered the scent of another stoat, as other stoats are competitors but not predators. Upon encountering the scent of predators–cats or ferrets–they quickly consumed food provided by the researchers. The results suggest that the scent of these larger predators reduced stoats' perceptions of the risk associated with encountering a member of the same species.

Australian swamp rats are eaten by feral cats, but less so by Tasmanian devils, which will kill cats. If the enemy of my enemy is my friend, then swamp rats should feel safer in areas with devils, whereas cats should feel less safe in areas with devils. Indeed, cats were detected by camera traps (movement-triggered cameras that wildlife biologists put out to census wildlife) less often in areas where devils were detected. If rats are sensitive to this relationship, they should be less frequently caught in live traps containing cat scent compared to devil scent. Results of one experiment were suggestive but not entirely conclusive. Swamp rats were less likely to be caught in traps scented with feral cats than those scented with devils, but they were also less likely to be caught in traps scented with herbivorous native marsupials (pademelons) compared with devils. More experiments are needed in this and other systems.

Taken together, the scent of danger that elicits fearful responses may be associated with the scent of death and digestion. These scents can be excreted by carnivorous mammals in their urine (PEA) and in their feces (TMT). They seem also to be present in blood (E2D). In addition, scents in predators' fur can discourage prey from occupying an area, and vulnerable prey species, sensitive to these chemicals, have evolved the ability to detect these scents. Wise parasites with complex life cycles may manipulate the abilities of prey to detect these chemicals to increase the likelihood that they, the parasites, will be transmitted to their ultimate host. Some predators have evolved olfactory cloaking abilities to reduce the likelihood of detection by their prey.

But what does danger smell like for humans? As we diverged from our mammalian relatives our olfactory system evolved and specialized.

How we perform on olfactory discrimination tasks is largely a function of the specific odors used. We are even better than rodents at detecting certain odors because either we've evolved the ability to detect them or retained the capacity to detect them that our ancestors had. But if humans have very few TAAR4 receptors (responsible for detecting PEA), and if humans don't have a vomeronasal organ, then is it possible that we don't have sophisticated abilities to detect the scent of danger?

We certainly respond with disgust to specific odors, including decomposing meat, feces, and vomit. But disgust is an entirely different emotional response from fear. Disgust keeps us away from undesirables, but fear keeps us away from predators.

Considering the research discussed in this chapter, we know that if animals or humans can smell or otherwise detect the scent of danger, then it is likely to be innate or quickly learned. And when learned, only a single traumatic experience will lead to fear of the odor's source. We may be primed to learn to associate scents with traumatic experiences, even though we may not detect them directly. I'll address this topic in Chapter 7.

It's best to dodge predators and their smells in the first place, whether by keeping them at a distance, becoming more wary, or avoiding their territory altogether. Indeed, managing exposure to predators through increased wariness or avoidance of risky areas is the most common way animals (and humans) avoid getting killed by them. In Chapter 5, I'll explore how animals steer clear of risk.

5

BE VERY AWARE

Herodotus, the fifth-century BC historian, wrote of fantastic animals during his travels in what is now northern Pakistan. "Here, in this desert [he wrote], there live amid the sand great ants, in size somewhat less than dogs, but bigger than foxes. The Persian king has a number of them, which have been caught by the hunters in the land whereof we are speaking. Those ants make their dwellings underground, and like the Greek ants, which they very much resemble in shape, throw up sand heaps as they burrow. Now the sand which they throw up is full of gold."

Years ago, while conducting a marmot study in northern Pakistan, I realized that Herodotus was writing about marmots, not ants! Herodotus saw gold washers sifting through diggings outside marmot burrows on the Dansar Plain, an area jointly claimed by India and Pakistan that overlooks the Indus River. Perhaps, I thought, studying golden marmots could be profitable in more ways than one!

In this chapter we will learn about where and when animals should be afraid, and what that implies about our own fears. We will learn that it's impossible to completely eliminate risk. We will see how knowledge of animal security may shed light on our aesthetic preferences. And we will see what happens when animals living without risks suddenly encounter them again.

My study aimed to quantify the risks that animals accept while engaged in different activities. Golden marmots were an ideal subject because they lived in a very intact predator community that included foxes, wolves, snow leopards, bears, and eagles, and because they are burrowing rodents. Animals that seek safety in their burrows are exposed to enhanced risks while away from their burrows. Although the distance to a burrow may be a reasonable metric, the time needed to get back to the burrow should be a better metric of risk. Travel time might be influenced by how heterogeneous an animal's habitat is. For example, if animals running uphill take longer to cover a given distance than animals running downhill, we might expect that risk does not increase in concentric circles around burrows but rather varies according to the time it takes to return to a burrow. For these reasons I studied factors that influence how fast animals run. The logic is that an animal running over an incline or rocks will run more slowly, and it will take longer to return to safety.

However, travel time is only one component of the total time to reach a burrow. Imagine a marmot focused on completing a really important task—one that will provide great benefits. A marmot could be engaged in chasing off an adversary that just wandered into its territory seeking to steal its food or to mate with its partner. For marmots, being able to gain sufficient mass to survive hibernation is essential, and that requires sufficient food. Additionally, since leaving descendants is the name of the game in evolution, protecting a valuable mate is also very important. For the marmot, all concentration is focused on chasing off the competitor. In these circumstances it's reasonable to think that the marmot may not be able to respond to other, potentially important, stimuli.

Indeed, cognitive psychologists have long viewed attention as something that is finite and divisible, as we saw from the hermit crab experiments in Chapter 2. If animals focus their attention too much on one thing, then there is no remaining attention to focus on other things. Studies on blue jays have shown this to be true. When intently foraging for cryptic prey, blue jays are less able to detect peripheral stimuli that might include predators.

If focused activities require different amounts of attention, the time it takes to respond to a potential threat while engaged in a specific activity could be another factor that influences the total time it would take an individual to return to the safety of its burrow. I studied this by broadcasting alarm calls to marmots who were engaged in a number of different behaviors, including playing, fighting, foraging, and looking. I found, not unexpectedly, that playing marmots either didn't respond to the broadcast alarm calls or took a relatively long time to respond. And, when this response time was added to the expected travel time, the playing marmots took a long time to return to the burrow. In contrast, foraging marmots responded in less than a second to a broadcast alarm call and bolted back to their burrows. Therefore, play can be inferred to be risky because of its attentional costs. Interestingly, however, marmots seem to compensate by playing very close to their burrows.

I found that the distance to their burrow was not the only thing that influenced marmots' risk. Anything that increased the time it would take to return to a location of safety influenced risk. Exposure is more than simply distance; it is time as well. This insight has considerable predictive value for how animals reduce risk.

Some animals rely on dense vegetative cover to provide them safety. But cover can be either protective or obstructive. It is often a place for animals to hide when they detect predators, reducing the exposure to some terrestrial and most aerial predators. Many hawks and eagles would damage their wings if they flew swiftly into a bush. But not all raptors are so handicapped. Accipiters—hawks that eat birds—have thin, pointed wings adapted for speed and for entry to tree cover when chasing prey. Once, while on a supply run in Gilgit, Pakistan, I watched an accipiter, who appeared seemingly out of nowhere and flew into a tall, dense privet hedge. Seconds later, it glided out with a house sparrow in its talons. Just moments before the sparrows were all singing and chirping loudly both inside and outside the privet. They likely did not see the predator coming; I certainly didn't. They also likely felt somewhat safe in the dense vegetation. After the attack, the sparrows were stunned into silence. It was at least ten minutes before the chirping and singing resumed.

The tammar wallabies and western grey kangaroos I've studied in Australia vary quite a bit in size: tammars are cat sized while the kangaroos are the size of an adult human. As we learned in Chapter 2, these largely nocturnal marsupials spend their day sheltering and snoozing in dense cover until, around sunset, they emerge from the cover to forage in open areas. My wife, Janice, and I spent many evenings observing, through image intensifiers, the wallabies and kangaroos foraging in places with and without predators. We found that at an area where both wallabies and kangaroos had a long history of predation by mammals and raptors, they varied in how they emerged from cover around sunset. Kangaroos cautiously observed the meadows for a while from the cover's edge before skipping out to the center, then they reared up on their hind legs and looked around. In contrast, the wallabies emerged slowly from the dense vegetation and foraged close to the edge. When alarmed, the kangaroos reared up and looked around, remaining in the open. The wallabies, however, looked like baseball players caught trying to steal second base; they hopped quickly back into the safety of dense cover. For wallabies, cover is protection, but for kangaroos, cover is obstructive.

Joel Brown, an ecologist, studied nocturnal desert rodents—pocket mice and kangaroo rats—that foraged on patches of seeds in the sand. As flowers dropped their seeds, the seeds and sand would blow around, creating little clumps. These clumps or patches, when buried, varied in how much food they would provide a hungry rodent. Profitable patches were those where the rodents were able to harvest a lot of seeds in a short period of time. Less profitable patches were those requiring more time to harvest the same number of seeds. As one would imagine, rodents preferred more profitable patches than less profitable patches. That is, rodents should prefer to forage in dense patches of seed and avoid the less dense patches of seed. Remember, rodents' nocturnal foraging time is limited. Thus they should be quite sensitive to patch profitability. If we assume that rodents that are sensitive to profitability will leave more descendants than those that are not, then rodents that forage optimally will be very sensitive to patch profitability.

Every individual will ultimately die. But let's assume that there are two ways to go: starve to death or be killed by predators. If an animal hides all the time to be safe from predators, it will starve to death. If an animal ignores predators and eats anywhere, it will have a greater risk of being killed. Let's return to Joel and his rodent experiments. Joel aimed to capitalize on the starvation-predation risk trade-off. He realized this powerful logic of trade-offs allows us to consider risk proactively, and so he designed an experiment to ask animals directly about the risk associated with a particular patch. If the density of seeds in different patches is identical, but some patches are either in risky locations or have cues of risk associated with them, differences in the amount of food the rodents leave will be attributable to the location or risky cues.

Thus, Joel invented the Giving Up Density technique. Its abbreviation, GUD, rhymes with "mud." To run this experiment, he mixed a measured amount of seeds with a measured amount of sand, shook the mixture well, and poured it into trays. He placed these trays either next to bushes or in the open and left the mixtures out overnight. The next day he sifted out the sand and weighed or counted the remaining seeds. The number of seeds left in a GUD tray after a night of foraging by pocket mice or kangaroo rats would reflect the risk at the tray's location. Lots of leftover seeds equals a risky patch; few seeds left means the location is not as risky. Joel found that pocket mice left more food in the open while kangaroo rats left more food close to the bush. This technique provided evidence that different desert rodents perceive cover as either obstructive (kangaroo rats) or protective (pocket mice); more seeds were left in areas that a species perceived as risky.

Since the initial experiments, the GUD technique has been modified for other animals, including diurnal squirrels, gerbils, marsupials, hyraxes, ungulates, and a variety of small carnivores. It requires animals to learn to forage at the tray, something that may take some cautious animals a long time to do. Researchers placed GUD trays on lines radiating out from the trees where gray squirrels reside. Squirrels left more food at distant trays and consumed more food on trays close to the trees. Thus squirrels, like wallabies and pocket mice, can be said to be more apprehensive when foraging farther from locations of safety.

To test risk in marmots, Raquel Monclús and Alexandra Anderson put out piles of horse food at distances of one, five, ten, and twenty meters from a number of yellow-bellied marmots' main burrows. Since marmots love horse food, we expected they would eat as much of it as they could. We found that once the marmots discovered the feeding station, their behavior depended on how far the stations were from their burrows. When marmots were farther from their burrows, they took longer to begin to forage and spent more time looking up. The piles of food twenty meters away were untouched, while all of the piles one meter away were completely consumed. As marmots moved away from their burrows, they became more fearful.

We know that marmots, who seek safety in their burrows, perceive cover as risky. Cover impairs their peripheral vision and lengthens the time it would take them to respond to a predator. Marmots rear up on their hind legs more often when they forage in dense and tall vegetation in order to look around or scan for predators. But is visibility responsible for this behavior?

My friend and colleague Peter Bednekoff joined me one summer at the Rocky Mountain Biological Laboratory to conduct an experiment. Peter is an authority on antipredator behavior who seamlessly integrates theory and empirical studies. In extremely elegant experiments he manipulated the ability of birds to look around their environment while feeding. Our goal was to do similar experiments with marmots.

To begin we designed a foraging box, one meter on each side. The bottom, top, and one side were open. We could place either opaque gray plastic walls or clear Plexiglas walls on three sides. We then placed the box next to a marmot burrow, put a handful of horse food in the box, and varied the peripheral visibility to see how the marmots foraged with impaired visibility. Did marmots look around more when visibility was impaired? No! Marmots looked less frequently and foraged more intently when surrounded by opaque plastic walls compared to clear Plexiglass walls. They also were more reluctant to enter the opaque box. They went in and out of the box and looked around frequently before and after bouts of foraging. Marmots, it seems, are quite sensitive to exposure and seemingly minimize their exposure to risk by foraging quickly and efficiently when they are unable to simultaneously monitor risk.

Constantly looking around for predators can be stressful. You have probably felt your own heart race during or following a time of heightened risk. Changes in heart rate are a sensitive measure of physiological stress and anxiety. By putting heart-rate monitors on free-living animals, researchers found that white-fronted geese increase their heart rate when vigilant, startled, and when engaged in aggressive or defensive interactions with other geese. Horses increase their heart rate after being presented with a novel object or a novel arena. And free-living Uinta ground squirrels have increased heart rates in unfamiliar areas compared with locations near their burrows.

We also know that dominance status, health, and other state-dependent factors might influence the risks that individuals are willing to take, and thus these variables may also influence habitat selection. For instance, subordinate animals may be unable to compete with more dominant animals for access to food. In a number of species of birds that eat small seeds, subordinate or younger individuals take more risks by foraging sooner after a predator has appeared. This is because dominants will wait longer to return to a risky area since they can monopolize resources whenever they arrive. Similarly, subordinates may find themselves foraging farther from protective cover or on the periphery of social groups. One study showed that environmental conditions also drive individuals to take greater risks. White-throated sparrows forage farther from protective cover on cold and cloudy days—times at which the birds require more resources to maintain body heat. These individuals are making, as one might say, the best of a bad situation—they must accept greater risks to find food, but also they are more likely to be killed by predators.

Although African cape buffalo are renowned for their unprovoked attacks on humans, they have predators too. Despite their size and fierce fighting abilities, 14 percent of bulls are killed each year by lions, according to one Tanzanian study. By recording the location of buffalo kills, researchers mapped predation risk and found that the probability of being killed by a lion is highest at the edge of dense vegetation. Moreover, they found that buffalo are safe only when they are on mudflats—areas with great visibility. Like kangaroos, buffalo clearly find cover obstructive. That makes it all the more surprising that buffalo are more likely to graze in dense, vegetated areas at night, when the risk of lion predation is even greater. Are buffalo oblivious to the enhanced risks? Perhaps, in the long run, the energetic benefits of foraging in these risky places outweighs costs—an interpretation somewhat hard to accept given the extremely high predation rate. More research is clearly warranted to explain this paradoxical finding in buffalo.

Based on the results from many studies, we know that risk assessment is very context specific; after all, subordinate or hungry animals take more risks to forage than dominant or satiated animals. Thus, to really understand whether animals are doing their best in the face of their constraints may require quite a bit of background knowledge about the individual, the species, its history of risk in a particular environment, and its evolutionary history. The decision rules we've evolved to keep us safe and enhance our fitness have evolved specifically to solve problems we've faced in the past.

When I became interested in perceptions of risk more broadly, I knew I had a lot to learn from Dick Coss. Now an emeritus professor at the University of California, Davis, Dick is a brilliant and deeply synthetic thinker who has been engaged in a variety of interdisciplinary studies of risk assessment and perception that combine neuroscience, animal behavior, and evolutionary psychology. As an evolutionary psychologist, Dick aims to interpret human behavior through a functional lens. By assuming that cognitive abilities evolve, he seeks insights from the study of our ancestors. Dick has a particularly interesting hypothesis, which I'll describe shortly, about variations in how humans flee and find shelter.

My work as an ethologist (including this book) is inspired by the entire tree of life, whereas most evolutionary psychologists often focus on our more recent ancestors—those living mostly in the past ten thousand years or so—as well as our hominid ancestors. From the point of view of most evolutionary psychologists, the human population was stable for many thousands of years, and the human traits we see today reflect selection from that ancestral environment.

This approach is not without its critics, and evolutionary psychology taken to untenable extremes can generate unscientific just-so stories. But the approach, when used properly, can instead generate testable hypotheses that can be evaluated with multiple lines of data collected in different cultures. If a specific hypothesis makes a series of predictions that are evaluated and supported by empirical data, then we have correlative support for the hypothesis. Better yet, data may refute a particular hypothesis, which is an effective way forward because it narrows the field. In other words, a "strong inference" approach to conducting science generates multiple hypotheses with multiple predictions. All are then evaluated with empirical data. If results are consistent with one but not the other hypotheses, then we have gained a possible explanation for a pattern in nature. Ultimately, however, science is a process, and as new hypotheses and new data are generated, treasured hypotheses may fall by the wayside.

For instance, the theory of plate tectonics took about fifty years to be fully accepted following Alfred Wegener's insight in the early twentieth century that continents moved around and that this continental drift was what could explain the distribution of fossils and geological features, as well as living animals and plants on Earth. Without a clear mechanism at the time, many geologists and geographers argued against the hypothesis. These scientists attributed geological changes to heating and cooling of the Earth's crust. Eventually, the weight of accumulating evidence was too much, and their views were refuted.

Dick, an accomplished artist, is interested in the evolutionary roots of our aesthetics. He wants to know why we find certain pictures more appealing than others. He is particularly interested in the role of trees in landscapes. In line with evolutionary psychology's focus on recent ancestors, Dick hypothesized that in our hominid past our female ancestors, who weighed less than males, spent more time in trees. Female *Australopithecus afarensis,* he suggested, were more likely to forage near, flee to, and sleep in trees. While there is sometimes a perception that boys are better climbers than girls, the data show otherwise. Girls climb more often in playgrounds, and young boys are more likely to get hurt falling from playground climbing structures. Thus, there seem to be consequences that track the long evolutionary history of differences between the human male and the human female need to climb.

Boys and girls also respond differently to questions about safe locations. Dick and a colleague showed children a series of photographs depicting different environments. They asked children which environment they would feel safe in if they were chased by a lion that escaped from a zoo. Boys were more likely to point at the photographs with large boulders, and girls were more likely to point at those with trees. In other experiments, when participants were asked more probing questions about the structure of the tree, photographs containing umbrella thorn trees—the acacias present on the East African savannahs—were most preferred. Taken together, these findings are consistent with our expectation, based on this evolutionary scenario, that sex differences in the history of habitat use can influence sex differences in perceptions of safety.

Further research found that images of trees contributed to both relaxation and fear. Dick's work, funded by NASA, showed that human subjects felt more comfortable when shown photographs of distant forests and savannah landscapes. However, while trees may afford protection from savannah-dwelling primates, there's a long history of fear associated with dense forests. Humans depend on our sense of sight, and in dense forests we cannot detect risks from afar. Indeed, there are numerous fairy tales about children getting lost in the forest and encountering monsters and predators. The hero of "Little Red Riding Hood" is, after all, a woodcutter. Some researchers hypothesize that our predilection for cutting down forests and taming landscapes may stem from our innate fear of obstructed views.

Like the cape buffalo described above, other animals may be similarly naïve about the risks posed by predators, and habitat selection may be part of this naïveté. As I will discuss in much more detail in Chapter 9, the eradication in the early part of the last century and subsequent reintroduction of wolves to the Greater Yellowstone Ecosystem in the latter part of the last century has provided numerous lessons about habitat selection and fear.

Conservation biologists often refer to "shifting baselines." They mean that for things that change slowly, early experiences establish your perception of what is natural. For instance, sea turtles were so abundant in past centuries that early European visitors to the Caribbean wrote about how they were found in "infinite numbers" and in "inexhaustible supplies." Now, sea turtles are slowly recovering from centuries of overexploitation, and they have not nearly recovered to their initial abundance. Thus, scientists with a more recent window on turtle abundance might think that turtle populations have recovered, but much more wildlife conservation work is required to achieve the historical ecological state when they were found in "infinite numbers."

For a naturalist in Yellowstone in the mid-twentieth century, "normal" reflected riverbanks devoid of much vegetation and reduced avian diversity near rivers. This was because beavers, elk, and moose browsed with abandon the vegetation in riparian corridors. Living without wolves, one of these species' major predators, has resulted in at least seventy years of relaxed selection. The reintroduction of wolves led to a remarkable restoration of streamside vegetation. Beavers were scared back into the water, and moose and elk browsed areas with greater visibility. The change was rather sudden. How could this happen? Joel Berger's work provided a mechanism.

I've learned a lot from Joel over the years, about both biology and how to be a mentor. His intensity, enthusiasm, and humor are infused in his creative experimental design and masochistic drive to strive to answer the difficult questions. He lived in a campervan for years while studying ungulates in North America and Southern Africa, and over the last few decades he's spent his winters in the dark north—Alaska, Canada, Siberia, Svalbard, and other places—studying antipredator behavior. Each day these species face a starvation-predation trade-off.

Joel's work has shown that in the Greater Yellowstone Ecosystem, a cow moose's single negative experience with a wolf resulted in a fearful response to predatory cues. To study this, Joel dressed up as a moose! By doing so he could get sufficiently close to unsuspecting animals before hurling bags of feces and urine at foraging moose or broadcasting predator vocalizations. Female moose (cows) had limited responses to these different predatory cues in areas where wolves were historically missing. However, once a cow had a single negative experience, like losing a calf to wolf predation, she immediately responded to predatory stimuli by looking around and was more likely to flee.

Behavioral responses like predator recognition were mirrored in the ecological response of the vegetation in Yellowstone National Park. The formerly foraged vegetation in riparian corridors regrew as beavers and ungulates, fearing death, used the areas less often. What happened next surprised biologists—birds resumed nesting in those areas (more on this in Chapter 9), which, in a real sense, restored the original ecological organization of this community that was present when Lewis and Clark explored the area in the early nineteenth century.

For predator-naïve individuals, encountering a predator for the first time can be hazardous. Populations may find themselves without predators for both natural and unnatural reasons. Perhaps populations are isolated on an island with few predators, or perhaps humans have shot, poisoned, and trapped most predators. Likewise, predator-naïve animals may find themselves exposed to predators for both natural and unnatural reasons. Predator ranges may naturally change over time, or humans may introduce predators (or transfer prey to areas with predators). Understanding whether animals can learn to recognize predators and avoid risky areas is a topic of great importance to conservation biologists.

After reviewing recent studies, my colleagues Andrea Griffin, Chris Evans, and I realized that very few researchers were exploring how to teach predator-naïve animals to respond appropriately to predators. As I related in Chapter 2, Andrea rolled a taxidermically mounted fox into view of a predator-naïve tammar wallaby and then, wearing a witches' costume, chased the wallaby with a net. Wallabies quickly learned to run and increase their vigilance in response to the fox because the presence of the fox predicted the aversive event. In later work Andrea showed that tammars could learn only about specific predators because she was unable to train them to respond to a taxidermically mounted goat. She also showed that social learning is important: tammars could learn to fear foxes by seeing the response of previously trained tammars to the fox. Most such predator-conditioning studies focus on teaching animals to respond to the sight, sound, or scent of predators—they teach predator recognition. What we don't know about is whether we can teach animals to avoid locations where encountering a predator is probable.

Minimizing exposure to threats by spending less time in risky areas is a good way to increase longevity. Wise animals learn, or have been selected, to spend time in relatively safe places whenever possible. Exposure to predators probably should be measured temporally—both the time spent in risky locations and the expected time to return to safety—as I showed with my studies of golden marmots in Pakistan. The US military knows this. During the US occupation of Iraq, military convoys and defense contractors sped through Baghdad from the safety of the Green Zone to the safety of the highly defended international airport to reduce their exposure to insurgents who had been attacking convoys outside the fortified areas. Like squirrels running from the safety of one tree to the next, the US military thus minimized the time spent in areas with heightened risk to increase their chances of survival. This strategy had the unfortunate side effect of creating discord among Baghdad residents. The speeding military vehicles sometimes plowed through intersections, causing car accidents and associated fatalities. While this exposure-minimization strategy was an adaptive short-term way to reduce risk, it may ultimately have decreased security because of the lost support from Iraqis.

Being aware of threats is essential if we are to avoid them. But, as we learned from golden marmots and jays, behaviors have attentional costs that may prevent animals (and us) from detecting them. Because it's better to avoid having to identify a predator in the first place, it's generally important to find safe areas and avoid risky areas. We've seen how safety is context dependent and how different species view cover as either protective (pocket mice and tammar wallabies) or obstructive (kangaroo rats and kangaroos). We've seen how visibility may influence risk assessment in marmots, but also how it may figure into what humans find desirable or attractive.

A growing literature shows that photographs of natural landscapes may enhance health and well-being in hospitals. Reports show that patients exposed to certain pictures have less pain, feel better, and are released from the hospital sooner than those exposed to plain walls. While the evidence also shows that colors can influence perceptions of anxiety (red and yellow seem to be less soothing than green and blue), patients exposed to landscape and nature scenes are more likely to benefit than those exposed to more abstract art. Future work could focus on the specific attributes of scenes that enhance positive health outcomes. As Dick Coss's work may indicate, photographs or paintings of open landscapes with climbable trees could engender feelings of safety and enhance healing.

Risk is unavoidable. Yet, all of our ancestors successfully managed risk and evolved a suite of condition-specific behavioral responses to do so. This highlights the general lesson from life—trade-offs are ubiquitous. In Chapter 6 we will adopt an economic approach: we'll quantify the perceived threats and look at the myriad of factors that influence these trade-offs. And, as we will discuss in Chapter 10, there are time-tested strategies that tell us when we should overestimate risks and be cautious or accept risks and go for it. To avoid all risks, of course, would mean starvation.

6

ECONOMIC LOGIC

In December 2000, Janice and I traveled to Queensland, Australia, to study antipredator behavior in rock wallabies and birds. The birds of the Atherton Tablelands are especially diverse, and I was excited to increase the number of unique species we studied. I had just started to develop a large comparative database of flight-initiation distance in birds, a subject not well studied at the time. Flight-initiation distance, or FID, is the distance at which an animal begins to flee an approaching object. Virtually all species respond fearfully to approaching humans, and there's now a very large literature on "flushing," or encouraging animals to flee from their current location. Janice and I were creating a large database of human-stimulated FIDs to answer the fundamental question of why some but not all species tolerate humans. By comparing various species' fearful responses, we aimed to understand how these relationships had evolved, creating what we describe as an ecology of fear.

Data collection was particularly enjoyable because we began by finding and identifying the local birds. Once we spotted a bird, we slowly walked toward it while counting measured paces, noted the location at which it looked around, the location at which it fled, and the location where it was when we began our approach. Our vigilance and diligence paid off: a fortnight later we left with a much larger data set and only a few scabs from the abundant terrestrial leeches. What did we learn from such apparently simple data? Escape decisions nicely illustrate the economic logic that underpins all decision making—whether by animals or by us.

How do animals assess risk? In this chapter we will explore the huge literature on FID to understand the dynamics of risk assessment. These lessons have important implications for ecotourism since animals respond fearfully to us, just as they respond fearfully to predators. By understanding the economics of fear we gain more insights into the trade-offs animals (and humans) make on a daily basis that enable them (and us) to live another day.

Animals that allocate only the minimum time, energy, and resources needed when responding to threats will outcompete and ultimately leave more descendants than those that overreact to fearful situations. The common assumption that animals should flee approaching predators as soon as they detect them may be wrong. In fact, Darwin, writing in what's now known as the *Voyage of the Beagle,* was puzzled by why he could get so close to animals on the Galápagos, musing about what later became known as island tameness. Since Darwin's observations, scientists, especially behavioral ecologists, look for rules that explain behavioral diversity—both within-species diversity and between-species diversity—in an explicitly functional and evolutionary context. Behaviors that have net benefits are selected and evolve; those that have net costs are eliminated by natural selection.

Ultimately, costs and benefits should be quantified in terms of fitness. Behavioral ecologists think of "fitness" a bit differently. We don't measure how many daily steps an individual takes but rather how many genes from an individual survive to populate future generations. After all, evolution is about having your genes survive at higher frequencies than other individuals' genes. Viewed this way, fit individuals are those whose genes are disproportionately represented in future generations, while unfit individuals are those not genetically represented in future generations.

Though theoretically straightforward, it's exceedingly difficult to properly measure evolutionary fitness in most situations. It's not so easy to follow genes into future generations in the field, nor to attribute a specific action to a fitness outcome. Thus, behavioral ecologists take a shortcut: they quantify things that *should* be correlated with fitness: time, energy, and opportunity. Viewed this way, a costly behavior is one that takes time or energy, or prevents an individual from doing something else that may offer better results. All else being equal, it's more costly to run than to walk and less costly to sit than to move. Expending energy is a bad thing, unless it's necessary for survival. Running or fleeing from predators, for example, is clearly worth the cost. Through use of these fitness surrogates and through FID experiments like the ones that Janice and I conducted in the Atherton Tablelands, we can identify rules that govern escape behavior. Let's take a walk through some of those rules, and I'll show you how they describe evolutionary fitness.

In 1986, Ron Ydenberg and Larry Dill, two researchers based at Simon Fraser University in suburban Vancouver, published a deceptively simple and remarkably elegant model of escape behavior. They realized that it was essential to consider not only the benefits of flight—fleeing immediately upon detecting a predator—but also its costs. If an individual flees too soon, an opportunity is lost. If an individual flees too late, the predator may be successful.

To illustrate this, they graphed the relationship between the distance to the predator and the cost of remaining, which is their term for the benefits of fleeing. At large distances, the cost of remaining is relatively low. It increases as the hypothetical predator gets closer and closer to the prey. Thus, if distance is the x-axis on a graph, the curve slopes downward as distance increases. On the same graph, they also plotted the relationship between predator distance and the cost of fleeing, which increases linearly. The cost of fleeing from a distant predator is much higher than the cost of fleeing from one nearby. Because the cost-of-remaining curve declines with distance while the cost of flight increases, there is a point where the lines cross, and this point represents the optimal distance, from the prey's point of view, for flight.

What sorts of things could influence the cost of fleeing? Wouldn't it be better to flee immediately if there is a predator nearby? Ron and

Larry suggested that lost foraging opportunities are one such cost. Envision two birds: one is foraging and the other is simply sitting and digesting between bouts of foraging. If foraging is at all limited, then the bird that is foraging would lose more by stopping its activity to flee an approaching human. Thus, it would have a greater cost of flight. Using economic logic, Ron and Larry noted that the optimal location to flee was farther away when the cost of flight increased more slowly (such as for a resting animal) and closer when the cost of flight increased more rapidly (such as for a foraging one). In other words, animals should flee when the benefits from fleeing exceed the costs of remaining.

More generally, behavioral ecologists consider behavior optimal when an individual selects the strategy that produces the greatest fitness when given a set of possible strategies, whether immediately fleeing when a predator is detected, or waiting ten seconds before fleeing, or fleeing when the benefits most exceed the costs. Since we've defined maximizing fitness as maximizing the benefit-to-cost ratio, this decision should, by definition and on average, be the optimal solution. We assume that animals who flee optimally will survive and allocate their energy into producing more descendants than individuals using the other possible strategies. Importantly, "optimal" is defined with respect to *possible* strategies—not the best conceivable, but rather the best in a set of possibilities. While it might be optimal for an impala to fly away from a pack of wild dogs, impala can't fly.

Once Ron and Larry's model introduced the idea that escape decisions should be optimized, many researchers began to study factors that influence escape behavior. Some studies found that animals in larger groups tolerated closer approaches before fleeing. Other studies found the opposite. Some studies found that when animals were in dense cover, they tolerated approaches to a closer distance. Other studies found the opposite. What accounts for this diversity of results in experimental studies? And with this diversity, is there a general principle that explains the many paths of escape behavior that we see in nature?

Many disciplines of scientific research borrow methods and ideas from other fields, and old problems are often solved with new tools and techniques. Behavioral ecology is one such discipline. We have had much success using ideas from other disciplines to help us identify general principles. Economic tools can help us understand animals' foraging decisions, and similar tools can give us insight into when it's profitable to defend a territory and when it's profitable to fight for access to resources.

Species share relationships with their close relatives. Dogs and wolves, which are closely related, are more similar to each other in their behavior than are dogs and cats, which are more distantly related. Closely related species may share body size, brain size, eye size, reproductive strategies, habitat preferences, and overall appearance. If these different traits influence escape behavior, then we'll want to know how to account for the similarity that is expected because some species in a given data set may be more closely related than others. This can be a problem whenever there are biases in how data are collected. If, for instance, it's very easy to collect data on different species of crows, we wouldn't want a disproportionate number of crows to influence our general conclusions about the effect of, for example, body mass or group size on escape behavior in all bird species. Evolutionary biologists and statisticians have developed a number of methods that permit us to remove the effects of these close phylogenetic relationships and truly isolate the relationship between body mass or group size on escape behavior. Yet different comparative studies might produce different answers because they included different species that were studied at different times and in different locations. After all, we expect the environment to factor into some of the differences in behavior, due to how species have adapted to their environment.

Fortunately, there is a statistical approach that permits us to combine the results of different studies and allows us to draw more general conclusions about the diversity of behaviors. Called meta-analysis, this approach, routinely used by biomedical researchers who want to draw robust conclusions about whether or not a particular therapy is effective, has been successfully adopted by behavioral ecologists. To note, meta-analysis is a statistical analysis of published results. Rather than simply saying that five species tolerate closer approaches when in larger groups while ten species flee sooner when in larger groups, a meta-analysis estimates the average effect size—the effect one variable has on another variable of interest—using all the evidence. It places more value in studies with more data and less value in studies with fewer data. By focusing on effect size, we understand the importance of a particular variable to explain the differences in an outcome—in our case, flight initiation distance.

Effect size, unlike traditional statistical significance, is not greatly influenced by the number of observations or amount of data you have. Rather, it's describing the standardized consequence of a variable or a treatment. Contrast the effect on your longevity of smoking one cigarette or putting a single bullet in your head (please don't do this at home!). You will survive much longer after smoking one cigarette because there is less uncertainty in the outcome of a bullet than a cigarette. This is not to say that heavy and regular smoking has no effect; it does. One estimate is that each cigarette takes eleven minutes off the lifespan of a heavy smoker. However, one bullet will most certainly ensure that you don't make it to the next day. The effect size on longevity of a bullet is much greater than a single cigarette.

Meta-analyses can allow us to estimate the relative importance of different things that might explain differences in how soon species flee. A meta-analysis will tell us how important group size or sex or brain size or eye size is in explaining the differences we see in flight initiation distance. Meta-analyses will, therefore, allow us to identify the *generally* important costs and benefits of flight.

Ron and Larry focused on the highly dynamic decisions made by individuals, not species. But there are also differences among species, especially regarding the distance of a potential predator before flight. For instance, it's much easier to walk up to a hummingbird than to an eagle. Why? What could explain these species-level differences? This was why Janice and I were walking toward different species of birds around the Tablelands. We were collecting a sufficiently large data set on FID from different species with very different life-history traits, including body size, typical group size, age at first reproduction, brain size, and longevity. All of these traits are likely to influence the ways that species allocate limited energy to survival, growth, and reproduction.

An insight gained from studying many species is that it's possible to describe individuals and species as having relatively fast or slow life histories. Do they reproduce early, have relatively more offspring, and die young? Or do they mature slowly, delay the onset of first reproduction, have relatively few offspring, and live relatively long lives? Decisions animals make about allocating energy determine these life histories. An animal taking risks by allocating energy to rapid early growth and reproduction has a higher likelihood of death. By contrast, a cautious individual will make decisions that might slow its growth but increase its safety and survival rate. These cautious individuals will allocate more energy to each of their offspring, resulting in enhanced survival.

In nature we see species and individuals adopting both strategies because, ultimately, both can be effective ways to leave descendants. It depends on the specific environmental risks and on an individual's future prospects. Individuals reared in dangerous environments poor in resources may ultimately do better by living short, risky lives and reproducing at a young age. Even if they are cautious in such an environment, they may not survive to live a long life.

My aim with our bird studies was to see how life-history traits may explain the differences in FID and thus how life history influences risk taking. I also included a number of natural-history traits, like habitat openness, to assess how the environment influences risk taking. Some habitats, like those of an Antarctic penguin, have good visibility, but others do not, like the dense forest where vireos live.

For my first comparative study I created a data set of 150 bird species with ten or more FID estimates for each species. I found that the initial distance of a human who walked toward an animal was the most important predictor of the distance the animal fled. If we began walking toward them from farther away, the birds were alerted but did not flee immediately.

This finding led me to propose the "flush early and avoid the rush" hypothesis—conveniently abbreviated FEAR. I hypothesized that animals that had to allocate their limited attention on monitoring our approach would begin to pay costs for this behavior at some point, and would benefit from moving away to reduce these monitoring costs. In other words, if animals were less able to eat, court mates, or monitor other potentially more important predators while keeping an eye on us, then they were paying a cost by monitoring us. Ultimately, those that fled early would not pay these costs and would allocate their time more efficiently. While not all species flee soon after detection, FEAR has been supported by both comparative analyses and meta-analyses. This means that FEAR is a generally important factor that influences escape decisions. Further, life-history and natural-history traits account for the differences in how close you can approach an animal before it flees.

After starting distance, the next most important variable explaining differences in FID is body size. Big birds, mammals, and lizards studied in places with few humans initiate escape sooner than smaller animals. But it is these very same large-bodied species that seem to habituate to humans if they are present and appear benign. We defined tolerance as the difference in FID at a place with many people and a place with fewer people. Thus, a tolerant species was one that let people approach much more closely in the area with many benign humans. The most important factor for tolerance of humans is the type of human activity in the study zone. We compared birds in urban areas and suburban areas, and those living inside and outside protected reserves, among other habitats. When animals were able to tolerate humans at all, the urban-rural contrast has the biggest effect; urbanization seems to make species more tolerant. Interestingly, the second most important variable explaining tolerance is body size. Large species that tolerate humans are the most adaptable.

Another factor that influences FID is predation risk. Birds who live in areas with more predators flee humans at greater distances. FID also decreases with sociality. More social species tolerate closer approaches, possibly because they live in larger groups. Additionally, body condition plays an important role in animals' escape decisions. Parasitized birds tolerate closer approaches, possibly because it is more costly for them to flee. Hunted birds, wisely, flee at greater distances.

Brain size also has a profound effect on escape behavior in birds: birds with larger brains tolerate closer approaches. While larger species of birds have larger brains, we accounted for this statistically and suspect that larger-brained birds can tolerate closer approaches because they are better able to assess and evaluate risks. Once a risk is detected, animals must allocate their limited attention to monitoring the approaching threat. If this monitoring is too costly, animals should flee to reduce the ongoing costs of monitoring, the basis of the FEAR hypothesis. But the brain is the organ that assesses risk and allocates attention, and larger-brained species have the cognitive abilities to monitor risk while engaged in other activities.

Geographic location or environment is an important variable. A number of studies have shown that birds living in the tropics have a greater risk of dying young from predation due to the greater number and variety of predators in comparison to higher latitudes. This parallels other latitudinal trends showing many more species in the tropics than at the poles. The greater predation risk has selected for a whole suite of life-history responses—tropical birds typically live faster, invest less in more offspring, but die sooner than species living in temperate regions.

A fascinating finding is that you can get closer to female but not male birds at higher latitudes. Males maintained wariness across latitudes while females were less wary at higher latitudes. This was not explained by differences in body size between males and females, nor was it explained by differences in their color (many male birds are more brightly colored than females). What could explain this geographic and sex-specific result?

We interpreted this data as a result of reduced predation risk for nestlings at higher latitudes. Liana Zanette and Michael Clinchy's results, presented in Chapter 3, were instrumental in this interpretation. Recall that birds hearing predator vocalizations reduced the number of feeding trips to their nests. If this is a generalizable idea, it makes concrete predictions about how risk should vary on an elevational gradient because altitude leads to the same patterns of predator diversity as latitude; there are fewer predators at high elevations and at higher latitudes. Thus, we may also expect fear, as measured by FID, to vary along an elevational gradient as well.

Lizard escape behavior is influenced by similar factors. Things that influence the cost of fleeing—such as food availability and social interactions—are the most important factors explaining differences in lizard escape decisions. Predator density is very important as well. We know that lizards and birds living with more predators are warier. As we expected from our understanding of avian fear, habitat factors also influence escape decisions, including how far lizards are from a refuge and whether or not they are in dense cover. Lizards far from their refuges escape at greater distances and those in dense cover remain there. Finally, the predator's behavior also influences escape. Predators moving quickly toward prey drive prey to flee sooner.

Can the insights from these studies help address my initial question about why there is variation in species' tolerance of people? As discussed, insights from FID studies have suggested some generalizations about the effects of body size. Large-bodied species in locations where they do not interact much with humans are more likely to be disturbed. Colleagues and I created a computer model that assumed that these distractions reduced the amount of time species could allocate to feeding and then looked at expected survival and reproduction. Disturbed animals are more likely to have reduced fitness when interrupted or agitated. However, this fitness cost drives selection for tolerance! Thus, large-bodied animals that are able to co-exist with people become more tolerant than smaller-bodied animals.

I wrote a draft of this chapter while on sabbatical in Sydney, Australia, where sacred ibis have become pest birds, better known as "bin chickens" in urban Sydney. Ibis stand almost two feet tall and have a long, curved bill. My favorite example of just how pesky they have become is an observation Janice made at Circular Quay in downtown Sydney. She sat watching a nearby ibis, perched behind a man reading his newspaper and eating his sandwich on a park bench. The ibis slowly inched its beak forward, around the man's shoulder, and with a swift, arching movement grabbed the sandwich out of the man's hand. Big birds. Big problems.

But what about the smaller birds? Do birds that become more tolerant of humans become more vulnerable to predators? My colleagues Diogo Samia, Benjamin Geffroy, Eduardo Bessa, and I tried to answer this question. In many cases predators avoid urban areas, leading to a "human shield" that protects urban prey from predation.

Human shields have been demonstrated to drive ecological and behavioral effects. They may trigger a cascade of events: predators avoid certain areas, and prey become more likely to frequent these areas. Prey then become less vigilant there because fewer predators hunt them. The reduction in vigilance means that prey can allocate more time to foraging, and the vegetation takes a noticeable hit. (We'll revisit this idea in Chapter 9.) But both human shields and increased exposure to humans may be present in other situations. Nature-based tourism is one notable example. According to a recent report, over 8 billion people visit terrestrial protected areas annually. It's as if each person on Earth visited a protected area once, and then some. And we know that there are deleterious consequences of this visitation: increased traffic and pollution, vegetation trampling, vehicular collisions with wildlife, and so on. Relatively small increases in mortality can tip populations from being seemingly stable to declining toward extinction. It would be ironic if the very eco- and nature-based tourism that seeks to protect nature is actually harming it.

I have always been a strong proponent of ecotourism. While studying marmots in Pakistan I wrote *An Ecotourist's Guide to Khunjerab*

National Park, a guidebook about the spectacularly beautiful high alpine national park in which I worked. I wrote it to help communicate with visitors and protect the park's natural resources. Well-managed ecotourism tries to minimize the ecological impact of visitors while maximizing their positive effects: cultural preservation, but also resources and income for those who protect nature. Because ecotourists should want to minimize any negative impacts, identifying them is essential.

If, as my colleagues and I suggested, too many nature-based tourists made wildlife more vulnerable to predators, it would strengthen the argument against ecotourism. We hypothesized that tourists created a human shield that protected animals, leading them to become more docile and less responsive to human presence. The human shield effect has been implicated in increasing vulnerability to wildlife poachers and illegal hunters. But would it also, under the right circumstances, increase vulnerability to natural predators?

At first blush the idea seemed unlikely. We know that many prey have sophisticated contextual cues—habitat, weather, or location—to assess predation risk and modify their behavior to reduce exposure to predators. In some cases, prey initially retain abilities to recognize "past ghosts" when colonizing predator-free areas, but these abilities may be lost rapidly if there are indeed no predators (as discussed in Chapter 2). In addition, we know that many prey species have the ability to discriminate among different predator species and react according to the degree of risk they face from that type of predator. Thus, the idea that by habituating to people, prey may not recognize another species of predator appears unlikely.

Nevertheless, the idea that individuals who become excessively bold around humans may also be bold around their predators has some empirical support, and we expect selection to package similar sorts of traits together. For instance, docile individuals may be less responsive to both conspecifics and predators. Thus, if by being around humans animals are generally becoming more docile, which has been reported in some species, like samango monkeys, then perhaps we are creating animals less able to deal with their natural predators.

So what have we learned about economic logic? Quantifying the costs and benefits of behavior helps us ask whether or not behaviors are adaptive. We have seen that a variety of factors influence the distance at which an individual flees an approaching threat, and we have also identified patterns in this flight-initiation distance that are species specific. FID is quite sensitive to costs, and we have learned that the costs of escape have profound impacts on the distances at which animals flee. If there are costs to leaving a good foraging patch, a preferred perch, or a promising social interaction, animals can be approached more closely; they tolerate more risk.

Because fear is the side-effect of an economic decision, benefits also influence the risks we will accept. As an example, my son and I like to watch big wave surf competitions. Neither of us want to surf sixty-foot monster waves. When they crash, these waves create earth tremors that can be measured on the Richter scale. The super athletes who choose to surf them train by carrying boulders and weights underwater while holding their breath. They must prepare for possibly being held underwater and dragged across the ocean floor for over a minute. For this risk, the successful surfers win substantial prize money. In 2017, the top award for the Mavericks Challenge Big Wave competition in California was $120,000. In Teahupoo, Tahiti, those who survive the very large (but not truly enormous) waves breaking on an extremely shallow coral reef can win over half a million dollars. There are benefits beyond the initial financial reward. Winners and others who have photogenic rides decorate the covers of the world's surf magazines and get lucrative sponsorship and advertising contracts. Thus, the benefits of successfully confronting fear in this scenario are obvious: great financial rewards. Would people stop surfing huge waves if the financial rewards were smaller? Probably not. After all, there is glory associated with mastering a huge wave and, at a very immediate level, the addiction of the rush.

Other decisions about accepting risk illustrate the economic arithmetic that we engage in when making decisions. I ask my students if they would go to the most dangerous place in the world, such as a war zone, if I paid them $100. Nobody accepts. However, as I up the ante they begin to squirm in their seats. $1000? $10,000? $100,000? $1,000,000? $10,000,000? For $10,000,000 they could afford a security detail, body armor, and other forms of protection. My students consider both the benefit of successfully confronting fears and the opportunity cost of not confronting their fears. We all have our price.

By understanding the economics of fear we gain insights into the trade-offs that both animals and humans make on a daily basis. These insights challenge us to seek strategies to successfully navigate them. But they also shed light on the effects of certain threatening environmental conditions in human society.

In humans, the logic of life-history theory has profound public health implications. If life itself is uncertain and resources are scarce, it makes evolutionary sense to reproduce early and often and invest fewer resources in each child. But this means the children themselves are at great risk. Those living in poverty would leave no descendants if they delayed reproducing too long, however, and died before they could do so. If a fetus or infant is stressed or experiences poor nutrition, the effects are seen throughout their lives. Research has shown that later in life, the pathologies of poverty include diabetes and heart disease, both of which are associated with reduced longevity. Ample resources and a stable environment provide the building blocks for maximal health in individuals and their descendants. We are thus descended from individuals who made the right decisions—those that maximized reproductive fitness—whether consciously or not.

Ultimately, however, we manage our risks and benefits, and this management is based on our perceptions of risks and rewards. We often get it wrong. Evaluating our perceptions is important work: it contributes to how we assess threats but also what will be needed to overcome irrational fears. US politicians who intend to instill fear by focusing on the rate of crimes committed by illegal immigrants often conveniently forget to note the higher rate of crimes committed by US citizens. Once people have made the association, however, it requires work to change their assessments to match reality.

Public perceptions of risk can also be influenced by a mistaken emphasis on a true but less relevant fact. Conservation biologists initially noted that the lionfish—accidentally introduced from the aquarium trade to the Caribbean—was poisonous, in addition to being highly invasive and destructive. This has inadvertently created problems for a creative and novel program to eliminate them—encouraging people to eat them! Potential consumers are thus wary of hunting or eating this really tasty fish, even though the risks of poisoning are eliminated if you handle them carefully with gloves. More work will be needed to overcome this manufactured fear.

Like the birds we've discussed in this chapter, we use economic logic to assess our risks, whether or not we are aware we are doing so. And for that we have to thank our ancestors. Their experiences prepared us to make the right decisions about how to maximize our benefits while reducing our costs in many situations. We are more cautious because of their experiences. A particular challenge arises when risks are truly novel. Consider identity theft—a truly novel risk with profound consequences for our security and well-being. We've not evolved the precise tools to be sufficiently skeptical of emails or phone calls that scare us into releasing our personal information. Novel threats, as we will learn in Chapter 9, require novel responses, and these may require some work. But for those who are anxious or fearful about threats, for the right price, these fears can be overcome. And this is good because we live in a world surrounded by risk.

7

ONCE BITTEN, TWICE SHY

Our then ten-year-old son, David, was in the lineup at Venice Beach. Most of the waves were in the six-foot range, with faces about twice his height. Sitting next to him on my board, I could see a hundred people dispersed along the beach, jockeying for the location where the best waves would begin to break. With luck, we could surf ahead of the break for about fifteen seconds before jumping off our boards or twisting back out to sea to catch another. David positioned himself correctly and paddled into a large, surfable wave. After a long drop down the face, he shot across ahead of the break before turning into the wave, leaping over it, and paddling back out. It was turning into one of his best days of surfing.

Then a large, more powerful set of waves came plowing toward the beach. Surfers began paddling hard to clear them so they would not get smashed by a wave or have to struggle through the whitewater. I paddled out as quickly as I could over the first of several eight-foot waves. David and many others were not as lucky. To my horror, he did not make it over the first wave and disappeared into the foaming white impact zone. As I bobbed up on the next large wave, I saw David in the foamy detritus of the previous wave. The leash that connected his ankle to the board had broken. Along with his board, he lost his ability to paddle quickly out of danger.

I bobbed up on the next wave and, with increasing dread, saw that David was in the same place. With each wave he was pushed to the bottom of the ocean, holding his breath while tumbling underwater like a rag doll. I could not safely help him. The bubbly whitewater was fully aerated, which made swimming and even treading water difficult. David and I both knew the secret to survival was getting out of the impact zone as fast as possible before the next wave. But David wasn't moving.

In moments that felt like hours, David eventually began to back paddle away from the impact zone. But then I lost him and saw nothing after the next wave. By the end of the set of waves, the ocean, from the beach to where the waves broke, was blanketed with thick white foam. Surfers were scattered along the foamy inside and began to paddle slowly back out to where they could catch waves.

I finally saw David on the beach, sitting quietly, his legs scrunched up to his body. Another surfer had paddled toward David, pulled him on his board, and brought him to the beach. He was fine, but shaken. "I almost drowned," he said.

I'm not certain David has recovered from that shock, even years later. Until that frightening day he'd only experienced the joys of surfing and had not fully realized the sea's immense power. Even though he was a strong swimmer and for years a competitive springboard and platform diver with the ability to confront his fears, the terror of trying to swim out of the highly aerated whitewater stayed with him. Six years later he still refused, perhaps wisely, to paddle out on big days at Venice Beach.

Once bitten, twice shy is an apt way of thinking about how and why we learn to fear threatening things. One fearful experience can have a profound influence on our perceptions of safety and well-being, as David's story so aptly shows, and we see evidence throughout the animal kingdom of individuals learning what to fear and what not to fear. Human fear can manifest as anything from mild anxiety to the symptoms associated with post-traumatic stress disorder. If we wish to understand how individuals learn to fear, we must realize that experience and context influence assessments of risk. We are constantly updating our assessments of risk with additional information.

Modifying behavior based on experience is an adaptive process that increases the number of descendants successful animals leave, which means it's often location- or context-specific and subject to trade-offs. We can learn based on our own experiences or on others', and social learning can be an amplifier that permits life-enhancing lessons to spread quickly and widely. Learning is an essential adaptation, one shared by many species. There are many insights to be gained by thinking about how experience can be used to modify behavior, whether in animals or in humans. Understanding how learning influences behavior is becoming more and more critical for planning how we will manage our ever-increasing interactions with wildlife.

This chapter will address a variety of issues associated with learning. I hope to provide you with tools and insights to better understand why we learn what we learn. In many cases, we learn to fear, which means we should be able to unlearn fear too.

Learning can be broadly defined as a process by which experiences change an individual's behavior over time. It's important to note that these behavioral changes should not be attributable to either time or developmental processes alone. For instance, we would not say that an animal has learned to fear a predator if an older animal runs faster from a predator just because it's older, stronger, and has longer legs. Likewise, we would not attribute David's avoidance of surfing Venice Beach to learning if avoiding water was part of the natural process of growing up for all humans. We would, however, say that learning has taken place if individuals exposed to predators flee faster than they did before exposure or if only those children who had almost drowned now avoid large surf.

Learning to fear requires one or more bad experiences. It may be a rapid process, or it may occur over time. For example, as discussed in Chapter 3, many species of fish are able to learn to associate a particular chemical cue with a predator the first time they encounter both together. This one-trial, rapid learning makes a lot of sense if you think about its function. David's response to almost drowning wired into his brain a fear of large, powerful waves at Venice Beach. This makes sense from an evolutionary standpoint: fearful stimuli should lead to rapid learning because individuals who learn to fear appropriate things will leave more descendants on average than those who take longer to learn to flee potentially fatal experiences. Natural selection has selected for organisms that do their best given the constraints that they face. A surprising trend has emerged from this process of natural selection: some animals appear to learn in a Bayesian way.

An eighteenth-century minister, philosopher, and statistician, Thomas Bayes, developed a logic of decision making based on accumulating evidence—learning. Differing from traditional statistical logic, Bayesian logic assumes that we have some prior knowledge about the likelihood of an event, and with experience we update our estimate. For example, a traditional statistical approach would assume that there is a fifty-fifty chance of being struck by lightning when venturing outside, which essentially means that being struck by lightning is random. By contrast, Bayesian logic assumes that on a sunny day there is much less than a 50 percent chance of getting struck by lightning, but if you are on a golf course or alpine peak during a thunderstorm, there is a much greater chance. Additional information—for instance, the distance between you and each successive lightning strike—improves the estimate. Bayesian logic learns from prior experience. Formally, one begins with what's called a *prior probability distribution* and updates the "prior" based on accumulated evidence. This new *posterior probability distribution* is, as Bayesians would assert, the best estimate of an event occurring—whether it's a lightning strike or an attack by an eagle or a terrorist.

Updating risk assessments based on new evidence should be ubiquitous. Some animals have been shown to behave as though they are using Bayesian logic. For these, we assume that natural selection has selected for reasonable priors. This is particularly true when we think about predation risk.

Animals living with snakes, for instance, might be more prone to respond alarmingly to a sudden encounter with a long, thin object than those living without snakes. Or dense, vegetated habitats may be avoided because obstructed visibility increases predation risk. When animals find themselves in novel environments—which occurs with increasing frequency when humans intentionally or unintentionally modify the habitat, or shift animal populations in a way that changes their landscape—we often see suboptimal outcomes that could be based on a now-faulty Bayesian prior. As an example, consider animals recently relocated to snake-free environments. If there are no snakes around, jumping back in response to a slightly curved stick that may look like a snake is an unnecessarily time-consuming response. Or, consider the customary benefits of not foraging near shrubs that may contain a sit-and-wait predator like a snake. If snakes are no longer present, avoiding shrubs unnecessarily eliminates access to potentially valuable food. We expect that, on average, animals making suboptimal foraging decisions will leave fewer descendants. More generally, we expect that natural selection will select for ways to eliminate costly responses if those costly responses are no longer needed.

The rate and conditions under which a species learns things is itself subject to natural selection and is the product of evolution. Tammar wallabies required only four exposures to Andrea Griffin in her witch's costume to learn to associate the fox with an aversive experience. They may have needed fewer exposures, but we wanted to be sure that they had a sufficient opportunity to learn, so we tested them after four exposures. Remember, tammars had some ability to respond fearfully to foxes without training; tammars were programmed to learn to increase their fearful responses to things that they had some fear of already. In contrast, they didn't think much of our taxidermic goat, which did not elicit a fearful response, and even four pairings of Andrea with the goat did not teach them to fear it.

Learning can be both rapid and location- or context-specific. The film *Rosewater* tells the true story of journalist Maziar Bahari, who was imprisoned by the Iranian government. Brutally interrogated over several months of detention while blindfolded, he could only identify his interrogator by his rosewater-enhanced scent. People who have been through experiences such as his often have traumatic flashbacks that are triggered by seemingly innocuous things—like the scent of rose-water—or other specific environmental features that were present during a traumatic event.

For individuals that encounter different sorts of predators, learning is an essential process that permits individuals to respond appropriately to salient stimuli, like an attacking eagle, while not responding to nonthreatening stimuli, like a falling leaf. But animals and humans can become overcome with alarm if they learn to fear the wrong things. For species that suffer only predictable risks, learning is potentially costly. For tammars, learning to fear a nonpredatory goat rather than or in addition to a predatory fox would waste a lot of time and energy.

In cases where risks and cues are predictable, and there is insufficient time to learn, we expect innate predator recognition abilities. A fascinating example of this is seen in Karen Warkentin's work with accelerated hatching in red-eyed tree frogs. Tree frogs lay their eggs in masses on tree branches, and until they hatch they are extremely vulnerable to snake predation. What's truly remarkable is that if a snake approaches an egg mass, the reasonably well-developed eggs have the ability to rapidly hatch (often in seconds, and on average much less than a minute), and the prematurely hatched tadpoles fall like rain from the egg mass. Karen has elegantly shown that the embryos react to specific vibrations that would normally be produced by a slithering snake approaching the eggs. Since there is no opportunity for learning from this sort of fatal attack, the embryos respond innately to these vibrational stimuli.

But many species work from some innate predator recognition abilities of scary sights, sounds, and smells, and further hone them with experience. Sometimes, these recognition templates permit the identification of novel species. We assumed that this was what permitted Kangaroo Island tammars to respond to European red foxes—a species with which they had absolutely no experience in their lifetimes and with which they did not evolve.

Some of the best contemporary behavioral ecological research on learning is conducted on fish. My friends and colleagues Maud Ferrari and Doug Chivers have carried out many of these experiments. Maud and Doug—a dynamic husband and wife duo—initially come across as opposites. Maud rapidly fires out an unending stream of incredibly novel and exceptionally clear ideas while Doug offers stellar logic and impressive data sets in a quiet, paced delivery. I always look forward to speaking with them at scientific conferences, where I learn from their both intensive and extensive work studying the nature of fear in aquatic systems.

The basic fish learning experiment involves holding a predatory fish in a tank until its associated chemicals become concentrated in the water. The water from this tank is then piped into a tank holding a prey species. An extract is made by grinding up bits of the prey species (often just the skin, but on occasion the whole animal), mixing it with water, and filtering it to remove small particles. It is then piped into the prey's holding tank. Boom! As soon as this skin extract, representing deceased prey, diffuses through the water, the prey learn that the predator's smell indicates danger. They respond fearfully when they detect the predator.

Working with larval damselfish on an island in Australia's Great Barrier Reef, Doug and Maud and their colleagues asked how the background level of risk influenced the ability to learn about nonthreatening things. As they note, learning about nonthreatening things is as important as learning about threatening things, yet it is studied less often. If learning is Bayesian, then we would expect that there are both cues of safety and cues of risk, but most studies focus on the risk. And if you're already in a risky environment, then the same sort of risk cue should mean different things about your probability of survival. After all, as discussed before, whether you're on an alpine peak or inside your home should change how you feel about lightning.

To study the effects of background levels of risk, Doug and Maud pre-exposed larval damselfish living with no risk and those living with some risk of predation (which they simulated by exposing them to predator odors over a series of days) to a novel, nonthreatening odor. The novel odor rapidly lost any salience in the fish exposed to no risk of predation. But the fish exposed to some risk of predation never learned that the novel scent was nonthreatening; they kept up their vigilance even to potentially benign stimuli because they lived in a high-risk environment. Background level of risk thus influences how animals learn what to fear and, conversely, what is safe. Organisms in risky environments are typically well attuned and, if anything, overestimate the risk of novel stimuli until more experience shows them to be benign. We will learn more about the logic underlying these biased assessments in Chapter 10.

Fish can be quite sophisticated in their ability to learn the specific level of risk that their predators create. For example, the concentration of skin extract should indicate the risk that a predator poses. Wise fish should infer that there is a greater risk when there is more skin extract in the water. And they have been shown to do this. Maud and Doug trained fathead minnows to learn that brown trout represented either a high risk or a low risk. They manipulated risk level by changing levels of skin extract, reasoning that more extract would be associated with either a closer predator or more predators. In the high-risk condition, the minnows responded fearfully by dashing around their tank and hiding in response to the brown trout, as well as by reducing their activity once sheltered. Sometimes the fish sought out other fish in the tank to shoal with. In addition to the direct response to the brown trout, the minnows also generalized this fearful response to a close relative of the brown trout—rainbow trout. What is fascinating is that they only made this generalization in the high-risk condition, presumably because they are very sensitive to particular risk cues—even if they are not perfect cues from a member of their own species. Tellingly, they did not generalize this fearful response to a distant relative of brown trout—yellow perch—under either the high-risk or low-risk condition. As species diverge over evolutionary time, they often smell less similar, and this limits the ability for one species to use other species as cues for predation risk.

We've found that learning transfers across similar sorts of predators. This is because predators often share similar scents, particularly if they're eating similar prey. Predators that hunt their prey in similar ways have converged on similar appearances, known as archetypes. Archetypes may be visual, acoustic, or olfactory. In the visual domain, the now-extinct thylacine, or marsupial wolf, with its long mouth, long legs, and wolflike body resembles a wolf or dog despite not being at all related to them because they hunt their prey in similar ways. Thylacines, like wolves, chase their food down (which requires long legs) and then immobilize it with their mouth (which requires a long and toothy snout). Acoustically, closely related predators like coyotes and wolves have similar vocalizations because of their shared ancestry. The archetype hypothesis is a way to show that prey recognize more than a single species and may even respond to novel predators.

Social learning from other members of their own species can act as a potent amplifier of fear. As part of her dissertation work, Andrea Griffin asked whether tammar wallabies could learn from others to become more fearful of foxes. After training a "demonstrator" wallaby to fear a fox (by the aforementioned taxidermic fox, net, and witches' hat), she paired the demonstrator with a naïve tammar—one who had no prior exposure to foxes. The demonstrator hopped away in alarm when the fox appeared. After only a few experiences, the formerly naïve tammar responded fearfully to the fox also.

But not all socially transmitted fear is adaptive. Humans sometimes experience hysterical contagion, which occurs when people either convince each other that something is the source of fear or they copy the behavior of others. A famous example of this occurred in Salem, Massachusetts, in 1692. It's thought that the adolescent girls, who began acting strangely and were accused of being witches, as well as the fearful members of Salem themselves represented a case of mass, socially transmitted hysteria. Not all such cases of hysterical contagion are associated with fear. A much more pleasant example is the Tanganyika laughter epidemic of 1962, involving uncontrolled laughing by schoolchildren. Although the children were not laughing constantly, it spread through the schools and may have taken a year to peter out.

In recent decades humans have started to transmit fear via technology. In 1994, during the Rwandan genocide, Hutu extremists used the radio to accelerate the spread of the massacre. They broadcast hate-filled messages and instructions to "kill the cockroaches"—the Tutsis. This social transmission encouraged more than 800,000 deaths in 100 days. In 2014, the Ebola epidemic's coverage on television and the Internet triggered socially transmitted hysteria. Fears were multiplied by our now twenty-four-hour news cycle, which favors maintaining viewer attention over providing reasoned analysis. For instance, Representative Duncan Hunter (R) from California suggested on the *Sean Hannity Show* that Islamic State terrorists who were infected with Ebola were attempting to cross into the United States through weak southern borders and infect the population with Ebola. This completely unfounded rumor was immediately refuted by US government officials. However, once unfounded rumors are created, they persist in perpetuity on the Internet for all to reread and spin to achieve their own outcomes.

Besides its potential as a force multiplier of fearful rumors, social transmission leads to desensitization to real or manufactured threats. Further to the American response to the Ebola outbreak in 2014, social transmission of fear exaggerated the likelihood of a breakout of the virus from cases being treated in hospital isolation wards. Or consider the US government's assertions in 2003, prior to the Iraq War, that Iraq had weapons of mass destruction. Through the mere repetition of incorrect assertions, people came to believe that Iraq had the capability to destroy the planet.

Indeed, there's evidence that repeating false messages can change people's perceptions of the truth, even when people have concrete knowledge that the message is incorrect—a phenomenon referred to as illusory truth. Ronald Reagan dubbed the Soviet Union the "Evil Empire" in 1983, George W. Bush referred to Iran, Iraq, and North Korea as the "Axis of Evil" in his 2002 State of the Union address, and in 2018, Donald Trump referred to many immigrants in a caravan of Central

American refugees marching toward the US border seeking asylum as "stone-cold criminals." When such statements are repeated and repeated, people's perceptions are changed whether or not there is supporting evidence for the claim, and many view these nations or groups more fearfully. Such perceptions have consequences for our foreign policy. As we will discuss in Chapters 10 and 12, there are reasons why we are particularly susceptible to messages that involve gory deaths caused by a virus, or biological or chemical weapons. But we should realize that we are particularly susceptible to messages that invoke fear, whether plausible or not.

If our assessments of risk emerge from Bayesian processes, however, we should be able to eliminate our fears. We've known one way that fear can be eliminated for at least 2,500 years.

> There once was a shepherd boy who was bored as he sat on the hillside watching the village sheep. To amuse himself he took a great breath and sang out, "Wolf! Wolf! The Wolf is chasing the sheep!"

Aesop, the fifth century BC storyteller, wrote long ago about the process of habituation, though he didn't use that name. Habituation leads to declined responsiveness to a stimulus, as well as its doppelganger—sensitization. As the fable teaches, the shepherd boy quickly loses the attention of the citizens once they realize his cries of wolf are unfounded. They learn that he is not truthful about the predator. So when a wolf appears, no one believes his calls, and no one comes to his defense.

While intensive research in the last century has led to well-supported generalizations about mechanisms of habituation, we have not yet developed a natural history of habituation. A natural history would help us predict how wildlife will respond to humans and anthropogenic stimuli. The need for predictive models has never been greater because our growing human population is urbanizing and increasingly seeking out encounters with wildlife.

As both urbanization and nature-based tourism expands, animals around the world are being exposed to humans. In 1950 only about 64 percent of Americans lived in urban areas, while in 2018, 82 percent did. In 2018 an estimated 55 percent of humans lived in urban areas, and according to the United Nations, by 2050 68 percent of the global population will do so. This rapid urbanization has had major consequences for the animals and plants that evolved in ecosystems with far fewer people. One strategy to preserve biodiversity is to set aside protected areas, such as parks, reserves, and wilderness areas, where human impacts will be reduced. Yet these protected areas receive an estimated 8 billion visits per year; a number that exceeds the number of people on Earth, and then some. Developing a fundamental understanding of how animals respond to humans is essential if we wish to preserve the life-sustaining biodiversity that surrounds us. As discussed in Chapter 6, animals initially respond to approaching humans as if they are predators, by fleeing and thus diverting time and energy away from other important activities, such as acquiring food, resting, or searching for predators.

Habituation requires repeated unthreatening exposures to a formerly fearful stimulus. To study the process of habituation it's essential to look at an individual's responses over time. If an animal habituates to repeated exposures to humans, it will, for instance, tolerate closer approaches before fleeing. Thus, if we find that flight initiation distances in an urban population are much shorter than those in a rural population, we would infer that the urban population is more tolerant of humans than the rural population. Habituation may explain this increased tolerance in urban areas, but other processes could also explain this resilience. For instance, the urban population may be composed entirely of human-tolerant animals. Or animals may sort themselves out according to their degree of tolerance: those individuals that are urban tolerant will be found in urban areas while those that are afraid of people will avoid urban areas. And last, it's also possible that natural selection has created urban-tolerant animals. Nevertheless, if we wish to understand whether habituation is occurring, we must follow individuals and track their responses over time and increased exposures.

Many urban populations of birds, mammals, and lizards allow humans to get closer to them than rural populations of the same species. We assume that this reflects repeated, benign contact and thus some degree of habituation. But this tolerance, while widespread, is not ubiquitous. Sometimes animals react in the opposite manner. Developing a natural history of habituation would help us better understand when animals are likely to become more sensitive to human presence.

Sensitization may be adaptive if it helps animals avoid potentially risky or costly situations. Elephants, remarkably, get quite upset when they hear bees buzzing around. Their response makes a lot of sense; elephants may avoid getting their sensitive trunks stung. Humans who are allergic to bees might react in a similar way. We are likely sensitized to sensational claims about Ebola, whether true or false, because we want to avoid getting horrific diseases.

But we should not always assume that sensitization is adaptive. For instance, drug addiction, in humans as well as in animals that serve as experimental models for drug addiction, involves sensitized responses. Experience with drugs like cocaine or methamphetamines increases the desire for more drugs. Worse for addicts is that the underlying neural circuits and neurochemistry involved in sensitization share many components with the underlying neural circuits associated with drug-seeking behavior. This suggests that the process of sensitization may not be adaptive in this situation but rather may prime individuals to seek out more drugs.

I used these insights about human visitors and our methods of studying flight initiation distance (FID) described in Chapter 6 to work with birds in Southern California. The habitats I studied differed based on the number of human visitors. In the course of my study, I noticed that out of fourteen species of California coastal chaparral birds studied, only four species had significantly different FIDs in those areas with relatively more human visitation than areas that were less frequented by humans. However, and somewhat unexpectedly, these four species fled us at *greater* distances when exposed to more people. The other ten species had no significant differences in FID as a function of quantifiable differences in human visitation. What could explain this pattern of apparent sensitization to humans?

It's important to note that this finding was in stark contrast to my studies of wetland birds in Southern California. The California coastal wetlands have mostly been filled in and turned into homes and businesses. The few wetlands that persist play a vital role in providing resting stops for migrating birds and wintering grounds for the snowbirds that have left the arctic winter for warmer, sunnier climes. On these wetlands, all of the species we studied were more tolerant of human approaches when they were routinely exposed to more humans, consistent with habituation.

I suggested that perhaps species living in limited habitats, such as the remnant wetland fragments in Southern California, may be more likely to habituate than those living in more contiguous habitats, such as the coastal chaparral, because those that live in remnant fragments may have already gone through a filtering process that eliminated less-tolerant species or individuals. The filtering process drove local extinctions of those species that simply could not tolerate humans. Thus, the only species that occupy these wetlands are those that are, to some degree, tolerant of humans. I called this idea the "contiguous habitat hypothesis." My hypothesis needs proper testing by evaluating it in other habitats, ones with more species and different types of disturbance. We still don't really know what precisely it is about humans that disturbs animals. Is it the number of encounters with pedestrians, people walking dogs, or vehicles? If so, is this a simple dose-response relationship whereby more people are more disturbing or is there a threshold that, once exceeded, causes disproportionately large impacts? Are they bothered by the smells or sounds or light associated with us? We know that all of these stimuli can have negative impacts on animals, but we need to develop a much better understanding of how, *precisely*, humans influence wildlife.

So we know that differences in historical exposure to predators as well as current exposure to predators may allow us to predict the degree to which a species will be prepared to learn about fearful things. And we suspect that having no options to leave a habitat fosters the ability to habituate, while having options to move off may foster the ability to sensitize to recurrent disturbances. As discussed in Chapter 6, large species are more likely to be disturbed by people, but large species that can coexist with humans are also more likely to become tolerant of them with repeated, benign exposures. Body size is an important life-history trait that is associated with other life-history traits. The rules animals have evolved that enable them to respond adaptively to potential threats are the result of past selection. The degree of match between current and historical threats should permit us to understand how and when species may (or may not) habituate to specific types of disturbance. Truly novel disturbances may be more difficult to habituate to than those that share features with other known threats. These generalizations mark the start of a more predictive understanding that can explain the conditions under which animals, and perhaps humans, learn to fear or learn to forget fearful stimuli.

Theo, our corgi, loved to bark at people passing through a parking lot next to our home. While squat, corgis are not really that small, and his bark was impressive and loud. Theo had spent time in Colorado, where afternoon thundershowers are common in July and August, but there is almost never thunder and lightning on the west side of Los Angeles, where we live. One day, while Theo was barking at people from our patio table, a thunderstorm struck our neighborhood with little warning. With the first clap of thunder Theo ran into the house, barking with fear and closely following Janice. Thunder is a very low-frequency sound, and it's loud, of course. I suspect Theo thought there was a much bigger animal outside, and he wisely feared it. We ended up wrapping his torso in a scarf to calm him down. Ever since this event he startles, alert with fear, when we're watching a movie that includes a sudden low-frequency rumble.

The fact that a single fear-inducing event can have lasting impacts on our sense of security, and seemingly Theo's, is most strikingly seen in post-traumatic stress disorder (PTSD). Exposure therapy, which can reduce PTSD symptoms, relies on desensitizing people to a formerly fear-inducing stimulus. But there may also be pharmacological therapies.

To understand how exposure therapy works and to develop potential drugs to treat PTSD, we need a model system. Much of the work on PTSD has used fear-conditioning studies in rats. When electric shocks are paired with a specific stimulus, rats quickly learn to avoid that stimulus. But PTSD is also characterized by nonspecific anxiety. My UCLA colleague Michael Fanslow has developed this anxiety in rats by shocking them without providing any predictive trigger. These anxious rats will then learn, with only a single trial, to respond fearfully when a stimulus is later provided. Studies have documented neurological changes in their fear-conditioning circuits that resemble the neurological changes associated with human PTSD. This creates a set of animals with which he and his colleagues are able to research pharmacological treatments to eliminate the disorder. Recent work in his lab has shown that by blocking stress hormone receptors in the brain, rats fail to become fear conditioned.

If this finding can be applied to humans, it may create therapies that reduce PTSD. I would certainly have considered taking something after my Kenyan attack to avoid permanently changing how I assess risks. Janice and I would have discussed giving David something to reduce his trauma after nearly drowning on Venice Beach. And for anyone who has been violently assaulted, such medicines may prevent years of trauma.

For those who already suffer from PTSD, the aforementioned exposure therapy is used to attempt to eliminate it. Recent work with rats has identified the possible neural basis of why it takes specific exposure to fear-inducing stimuli to erase fearful memories. Neurons in the dentate gyrus, located in the temporal lobe, are associated with both memory formation and memory extinction. Therapeutically, the idea is that by desensitizing people to the fear-inducing triggers, they will no longer suffer from the debilitating fears and paralysis that often characterize the disorder. If, for instance, someone's PTSD was triggered by an attack they experienced in a vehicle, they may have learned to associate being in a car with an increased likelihood of an attack. Therapists work with patients to repeatedly create safe experiences of being in cars. With enough support from their therapists and lots of practice, this prolonged exposure therapy may eliminate the fear-inducing PTSD triggers.

Fearful responses have been honed by natural selection to ensure our safety. We share an evolved ability to form traumatic memories from traumatic experiences, which for many species is essential for survival. Fish learn quickly about predatory threats, and tammar wallabies know they should fear predators that look like foxes, but not those that look like herbivorous goats. But exposure to new threats—dangers that our ancestors have not experienced—creates new challenges.

We all are prepared to respond adaptively to new challenges because we (as well as many of our ancestors) are able to modify our behavior based on experiences. We, and other species, can learn from each other, and this acts to increase the rate at which knowledge can spread through a population. But the lessons from life show us that it's not simply adaptive in many situations to learn; we must properly match our assessments of risk with reality. Learning is Bayesian. At a more proximate level, desensitization and habituation provide mechanisms to reverse traumatic events. And insights from these mechanisms provide the potential tools to enhance our coexistence with wildlife in an increasingly urban world.

8

LISTENING TO SIGNALERS

Ah-AAAAAH! Ah-AAAAAH! The peacock's alarm call pierced the oppressive premonsoon heat at midday in Jim Corbett National Park, one of the remaining tiger reserves in northern India. My friend Nancy and I saw many types of birds and played with an orphaned elephant baby that was being hand-reared. We'd ridden on domesticated elephants that provided a sufficiently high vantage point to look down into the ten-foot-high vegetation—mostly wild *Cannabis*—that covered the unforested parts of the park. We'd found tiger kills and staked them out. We'd seen fresh tiger footprints, larger than my hand, and looked up and around with some trepidation. But after three days we still hadn't seen any tigers. Then, on the fourth day we walked to a viewing tower near the campground and climbed up to the shaded platform to look for wildlife along a river that bisected the park.

Ah-AAAAAH! Peacock calls usually signal something they find disturbing, like the presence of an elephant, a boar, or possibly a tiger. The peacock flew noisily away. Minutes later, the riverbank was filled with monkeys—rhesus macaques—who looked back and forth before swimming quickly to the other side of the river. Once across the river, they climbed bushes and screamed while looking nervously back to the bank. A few moments later, a huge tiger emerged from the dense vegetation and slowly sauntered to the water's edge. Pausing a moment, the tiger waded in and stretched out for a cooling soak. The tiger was enormous, at least twelve feet long. All who could fall prey to it gave it a wide berth. I was fascinated that day back in 1987 by how the warning calls of one species influenced others. How common is such interspecific communication? And if animals routinely warn other species, what happens when we begin losing species from a historically stable ecosystem?

In this chapter we will learn about how animals gain information about predators through vocalizations; both from members of their own species and from other species. We will learn what modulates the production of alarm vocalizations and what they may mean, which has implications for the evolution of human language. We'll think about the reliability of alarm calls and the implications for others. Finally, we'll apply this knowledge about alarm communication to better understand how we might effectively acquire information and respond to threats.

If you don't have an opportunity to visit Corbett Park and hear the peacock's alarm, perhaps take a stroll through the woods or another natural area with a dog. Depending upon where you are, you'll likely hear the chattering of squirrels, the snorting of deer, and the trilling of birds. These vocalizations may serve three purposes: to communicate to predators, to communicate to other members of the same species (which we call "conspecifics"), and to communicate to other vulnerable heterospecifics (other prey species). While a vocalization may be intended for one purpose, targets are not mutually exclusive; a caller could be directing calls to all of them. It's possible to warn your relatives that a predator is around while also signaling to a stealthy predator that it's been detected. One of the many things that fascinates me about studying antipredator communication is that different recipients may select for varied features of alarm calls. To understand how, we first need to consider what we mean by communication.

Communication occurs when a caller produces a signal designed by natural selection to influence the behavior of a receiver. While animals respond to all sorts of cues for risk, as the monkeys did after hearing the peacock's alarm call, communication requires the evolution of specific signals that have been selected for. These signals change the behavior of others. But therein lies a paradox; why emit signals that advertise both your presence and your location to a threatening predator?

Much of the scientific literature on alarm communication focuses on how these calls function to warn members of the caller's own species. Specifically, a caller may benefit if relatives are warned. Recall the aim of evolution: to ensure that genes are passed on to the next generation. Because relatives share genes, the immediate answer to why prey would emit a potentially risky alarm call is that relatives are warned. By warning relatives, a caller protects its genes.

Classic research has shown that Belding's ground squirrels increase their probability of emitting an alarm call in response to a predator based on the specific composition of their audience. When more of their close relatives are present, squirrels are more likely to emit calls. When distant relatives are present, squirrels are less likely to warn. And when only nonrelatives are present, the ground squirrels are least likely to sound the alarm. But it can be a bit more complex than this; all relatives may not be treated equally. For instance, we have found that yellow-bellied marmots are most sensitive to the presence of their vulnerable young. Mothers with pups present are most likely to emit warning calls, while other individuals do not. In sum, one reason animals emit alarm calls is to ensure that their genes make it to the next generation, either directly (through their direct descendants—their offspring and grand-offspring) or indirectly (through their siblings, cousins, nieces, and nephews). Regardless of the specifics, the family that calls together stays together.

Of course, another reason to emit an alarm call is to create pandemonium. Imagine you are being stalked by a predator. If you sound the alarm, other members of your own species will scatter, and the sudden, rapid, and chaotic movement around you may both shield you from the predator and make it substantially more difficult for the predator to focus on you. Admittedly this response is a bit more self-centered, but it still achieves the evolutionary purpose of survival—at least, survival long enough to pass on genes to the next generation.

Finally, calls can be directed at other species. Recall in Chapter 3 we discussed Gunther's dik-diks, the gracile ungulates that are eaten by about thirty-six species of predators on the savannahs of Kenya. A student project studied whether dik-diks respond to the alarm calls from white-bellied go-away birds, a true sentinel of the savannah. They are large birds that perch on acacia trees and emit alarm calls when they detect predators. It's currently unknown why they provide this public service, but many species seem to respond to their calls. We aimed to find out if dik-diks were among the species that responded to the white-bellied go-away birds. By broadcasting go-away bird alarm calls to dik-diks and contrasting the dik-dik's response to nonthreatening bird song, we found that when dik-diks heard go-away bird alarm calls, they responded by running to cover, looking around more often, and foraging less. Thus a small, fearful ungulate can respond to alarm vocalizations from a bird.

Such eavesdropping is commonly seen in a variety of species and may be one of the benefits to species from living in what are referred to as "mixed-species groups." In most cases it's likely that the caller is directing its vocalization to either the predator or to its conspecifics, and that the eavesdroppers learn to respond to others' alarm calls. For instance, the diminutive but colorful Australian superb fairy wren has been shown to learn to recognize previously uninformative sounds as alarm calls. My Australian colleague Rob Magrath and his students trained fairy wrens to specifically respond to these sounds. They did so by broadcasting them through a hidden speaker and immediately showing the birds a model raptor. After two days of training, the fairy wrens responded to the sounds as they did to their alarm calls; they had learned to associate a novel sound with the information that it conveyed about the presence of a predator.

But what actually happens when an animal hears an alarm call? Depending on how deep down you go into an individual's physiological responses, hearing an alarm call can cause different genes to copy themselves, change the levels of stress hormones like catecholamines circulating in the bloodstream, or cause an animal to stop its current behavior and look around or flee, as we've just seen with dik-diks. We now know a lot about the conditions under which animals produce these calls, the physiological correlates with alarm calling, and the meaning of these calls. Based on my own work on marmots as well as key examples from research findings from other animals, including monkeys, meerkats, and rodents, we have learned a great deal about the immediate causes, meaning, and evolution of alarm calls. Let's begin our study with marmots—specifically, at the end of the marmot's digestive tract.

The marmots I study are herbivorous. Thus, they must eat a lot of vegetation, digest it a bit, eliminate it, and then eat some more. They spend their summer days eating, resting and digesting, and defecating. We, on the other hand, spend our days watching and live-trapping the marmots. Thus, it's not unexpected to walk up to a marmot resting quietly in its cage trap and be greeted by a fecal sample waiting to be collected. If we're lucky we collect even more when we reach into the conical handling bags we use to keep the marmots relaxed and still while we make measurements and collect data.

Feces are full of all sorts of good stuff to study. We can identify what marmots eat by looking at plant cell structure under a microscope, and we can find a variety of intestinal parasites by their eggs, which float to the top of a fecal slurry. Some folks extract DNA from sloughed-off intestinal cells in feces and use this to identify species and even count individuals of cryptic (difficult to see) carnivores. Feces also contain the digested remains of hormones—including stress hormones—such as the cortisol and corticosterone that we discussed in Chapter 1.

Quantifying these fecal glucocorticoid metabolites (FGMs) is a relatively noninvasive way to study physiological stress in animals. We've conducted a number of studies in our marmots that used FGMs. We know that we are looking at stress hormones in the feces because we conducted a simple experiment with a colleague's captive marmots. First we injected adrenocorticotropic hormone (ACTH), a hormone that stimulates the anterior pituitary gland to produce stress hormones. Then we waited and collected all the feces from the captive animals, noting the time elapsed from the ACTH injection. About twenty-four hours later we found a peak in excreted glucocorticoid metabolites. Thus, we infer that fecal samples tell us something about an individual's stress levels the day before we collected the fecal sample. While most individuals vary considerably both in the magnitude of and variation in stress hormone levels measured across multiple samples, some animals show multiple high-stress fecal samples. They are considered to have chronic stress.

Capitalizing on this knowledge, we began to note whether or not an adult female emitted an alarm call when we trapped her, and, if she was willing, we collected a fecal sample. Using this set of paired observations, we wanted to know whether a female was more likely to call when she had higher glucocorticoid levels. We discovered, rather unsurprisingly, that when females had higher background levels of circulating stress hormones, they were more likely to emit alarm calls. Therefore, stress hormones can be said to potentiate alarm calls, a finding that has been shown in a few other species, including rhesus macaques, a nonhuman primate. In the same way that you might be more likely to shout when you are under stress or scream when you are wandering through a haunted house, marmots and monkeys with higher levels of stress hormones are more likely to emit alarm calls.

Further, more recent work with our marmots at the Rocky Mountain Biological Laboratory has shown that socially isolated individuals are more likely to emit alarm calls. To determine this, we looked at individual marmots' position in their social network. You could say that marmots, like many humans, have Facebook profiles: they interact with different numbers of other marmots and have different types of social relationships. Perhaps, like the employees at certain social media sites, we quantified marmot social interactions and calculated formal social network statistics to snoop on our clients. We described how socially connected each individual marmot was with other individuals in a given year. By doing so over many years, we found that those marmots who were less popular, measured by fewer interactions with other marmots, and those that had weaker relationships, interacting relatively less often with others, were more likely to emit calls when approached.

These findings match another series of results suggesting that socially isolated marmots are uniquely vulnerable. Like marmots, we might be more vulnerable to being mugged when among people we don't know versus among a group of friends. We assume that our good friends will protect us if we are attacked, or that others won't attack a group. Our marmot results also suggest that vulnerable individuals might direct their calls either to the predator to discourage pursuit, perhaps, or to other marmots to gain status by informing them of the threat. More work is needed to understand this intriguing possibility of status signaling.

But what do the calls mean? Unlike birdsong, which communicates identity and presence, and may provide information about the singer's quality, alarm calls, and to some extent food calls, uniquely have the ability to be referential. Referential signals are signals that can refer to external objects or events. Thus, one type of call could tell others that there is a predatory fox around, while a different call would signal the presence of a hunting eagle. If so, alarm calls could function as basic words, and animals could use this information to develop a rich understanding of the world around them.

To understand why the study of referential signaling in animals may be a big thing, let's go back to Darwin. Darwin noted that while humans had language, nonhumans were able to communicate only their emotions and not information about specific objects or events outside their body. Thus, nonhumans were not expected to have referential signals. Later, when we learned that honeybee waggle dances can encode the direction and distance from the hive to a patch of flowers and communicate this information to other foraging bees, people at first were not too fussed—it's an exception to the rule in one species of tiny insects. Referential signaling in nonhumans challenges our belief that human language is unique—or at least it challenges the idea that only humans have the ability to communicate about external objects or events.

To test this potential, the quest for wordlike communication led to the development of a subdiscipline of researchers studying the meaning of the calls of primates, rodents, and birds. Evidence for wordlike communication, what has been called referential signaling, would be seen if two criteria were met. First, there had to be a high degree of production specificity; each call type was reliably produced by a signaler in response to the specific stimulus. For instance, upon detecting a fox, animals would always produce a specific-sounding fox-elicited alarm call. This call needed to sound different from the call elicited by an eagle. If, by contrast, a fox was detected very close to the caller and elicited a more eaglelike alarm call, the calls might be construed to be communicating the degree of risk and urgency of the response. In other words, the close presence of a fox is as risky as an eagle, which can fly swiftly and strike like a bolt of lightning. The second criteria was that there had to be some degree of contextual independence, or what's been referred to as "response specificity." For example, an animal hearing a fox-elicited call should respond as though a fox were nearby, while an animal hearing a raptor-elicited call should respond as though there was a raptor present. For animals with uniquely different escape strategies this is, in theory, relatively straightforward to quantify.

One of the earliest examples of such referential signaling came from a series of studies by Thomas Struhsaker, Peter Marler, Dorothy Cheney, and her husband and research partner Robert Seyfarth. Struhsaker and Marler's initial observations of vervet monkeys in Uganda led to a series of detailed experiments later conducted in Kenya. These cat-sized, black-faced monkeys live in the savannahs and seek protection in trees. Vervet monkeys may be recognizable even if their name is not immediately familiar. Males are notable for their quite distinctive brilliant blue testes and bright red penis. Females are notable because they inherit their social rank from their mothers; a rather depressing example of transgenerational inheritance. Their initial observations in Uganda suggested that the vervet monkeys produced and responded in different ways to the uniquely distinctive alarm calls elicited by raptors, snakes, and terrestrial mammalian predators—in ways that suggested that these alarm calls functioned as simple words that communicated to others the type of predator detected.

Dorothy, Robert, and Peter aimed to see if a Kenyan population of vervet monkeys showed the same ability. Their experiments are summarized in Dorothy and Robert's classic book *How Monkeys See the World*. Vervet monkeys spend a lot of time each day foraging on the ground, which is risky because the savannah is filled with predators, and trees provide a quick refuge. But vervets must travel between trees, and during this time they are quite exposed. They have evolved a variety of adaptations to reduce these risks: they can live in large, female-dominated social groups, which allows for increased security, and, of course, they emit alarm calls. The researchers found that vervets in the Kenyan savannah show an ability for referential signaling.

Upon detecting a python, vervets chutter. Upon detecting a leopard, vervets emit a series of short tonal calls. Upon detecting an eagle, vervets emit a grunt. Further, vervets hearing these predator-specific calls respond in unique ways. Their behavior changes based on the call. Upon hearing chutters, emitted once a vervet sees a snake, vervets get up on their hind legs and bipedally approach the caller, looking in the vegetation for a snake. Then, if they see the snake, they circle around it and harass it with chutters. They may even attack. Letting a snake know it has been detected and trying to get it to move away is a form of mobbing that many species engage in with predators that particularly rely on stealth for their success. Upon hearing leopard alarm calls, vervets run to trees and move to the most peripheral branches, where they might be safe from a heavier leopard, which can't climb out to those distal locations. And, finally, upon hearing the raptor-elicited grunts, vervets on the ground run to trees, aiming for the center of the crown to be safe from raptors who can't dive into the dense area without risking personal injury. Upon hearing the call, the vervets in trees drop down to the tree's center for safety.

Seeing vervets or other monkeys respond to a predator is quite exciting. I first experienced this in 1986 when I was studying blue monkeys in the Kakamega forest in western Kenya. Whenever a martial eagle flew over, it literally rained monkeys! Each monkey either jumped or rapidly climbed down from high branches to seek the safety of the central part of the tree, where the eagle could not easily reach them. More recently, in Costa Rica, I heard but sadly did not see (due to the lack of visibility) a troop of spider monkeys respond to what must have been a large predatory cat. They all moved to a single location and collectively produced loud, rapidly repeated vocalizations. The rain forest was filled with these cries, and they seemed to work. Eventually it sounded as though the now-detected predator slinked away in search of less-alert prey.

Dorothy and Robert conducted further experiments in the Kenyan savannah specifically designed to determine whether the monkeys responded in unique ways to the unique alarm calls. After they hid the speakers, they played back specific calls and filmed the vervets' responses. Through these broadcasts of predator-elicited vocalizations, they discovered a degree of response specificity that suggested the monkeys had simple words for predators. Thus, vervet monkeys can be said to produce referential alarm calls. The calls act as though they communicate predator type. Similar sorts of referential signaling have also been reported in other monkeys, in chickens and some species of birds, in Gunnison prairie dogs, and in meerkats.

But not all species have these referential abilities, and there may be some variation in this fascinating cognitive ability even among populations of the same species. A recent study with a South African population of vervets has failed to replicate Dorothy and Robert's initial findings. In the South African population, vervets did not immediately flee in predator-appropriate ways. Instead, they looked toward the hidden speaker broadcasting the alarm calls. When they did flee, it was not always with the expected predator-specific response. The authors—Nicholas Ducheminsky, Peter Henzi, and Louise Barrett—suggested this may reflect the substantially larger group sizes of these vervets and the greater distance at which individuals heard natural alarm calls. Thus, they proposed that the referential responses that Cheney and

Seyfarth discovered may not be a property of a species or a fundamental cognitive ability. Rather, referential responses may be more malleable and reflect social and ecological conditions.

In the case of the South African vervets, the calls from a greater distance implied vervets hearing them were at relatively less risk of predation. Much as you might immediately get down if someone next to you yelled "DUCK!," you might look around for the source of the risk if someone far away yelled "duck." The larger vervet group size implied vervets may have been generally more safe than those in Kenya. If so, we can hypothesize that an individual's relative safety may be an important factor that influences cognitive abilities. Animals living under relatively higher risk may need to be more cognitively sophisticated than those living under lower risks. Finally, in larger and more spread-out groups, there may be more false alarms. Vervets hearing false alarms should be more discerning and look first to identify the true threat before engaging in any predator-specific escape behavior.

Others have questioned the utility of studying the meaning of calls, and whether information is a valuable metric to quantify when studying communication. On the one hand, referential calls may not necessarily resemble words that label predators; calls could be instructions for others, such as "run to a tree" or "get up on your toes and look around." On the other hand, those that question the value of information contained in animal signals often ask what's in it for the signaler to provide meaningful information if communication is about manipulating the behavior of others. Besides, as these people note, information isn't an entity; it's an abstract concept. Evolution shouldn't act to optimize information transmission; it should act to optimize fitness.

My view is that signals contain potential information, and it's vital for animals to acquire information about potential risks. Even if it's difficult to imagine what information is precisely, you know it when you see it. If your behavior changes as a function of a detecting a signal or having an experience, you have obtained information about something that is potentially valuable. This is true for nonhumans too—if their behavior changes after hearing a vocalization, sniffing a scent, or seeing a particular image, those stimuli contained information.

I also find the quest for wordlike communication worthy of study, even if we find that it's relatively restricted. This is because it's an attribute of communication that is not found in all species, and this variability demands explanation. What explains the evolution of cognitive abilities? Why are some species "smarter" (admittedly a loaded word) than others? Ultimately, I think we should be studying the conditions under which communication referentially evolves or is favored by natural selection.

I've spent over a decade studying the evolution of referential communication in eight of the fifteen species of marmots. Based on previous reports suggesting that some species of marmots had a high degree of production specificity (seen when a single predator type elicits a single type of alarm call and different predator types elicit different types of alarm calls) while other species were not reported to have this ability, I set out with microphones, tape recorders, and speakers in tow to study both production and response specificity in marmots. And, ultimately, I aimed to study the evolution of referential communication in them as well.

My studies began in Pakistan, with golden marmots, then to the Alpine marmots in the Berchtesgaden Alps in Germany, to Ohio and Kansas for research on groundhogs, and Utah and Colorado to learn about yellow-bellied marmots. From there I traveled to the Olympic Peninsula and Mount Rainier National Park in Washington for my work studying Olympic and hoary marmots, respectively, to the steppes of Russia for work with steppe marmots, and on to the central mountains on Vancouver Island for research on the critically endangered Vancouver Island marmot. My wife, Janice, and I spent thousands of hours watching marmots in lovely Alpine settings and conducted a variety of experiments. We walked toward the marmots to observe their reaction. We flew an eagle-sized radio-controlled glider above them. We named our glider Eagle Knievel because it was difficult to land a glider on rocky hillsides. I spent hours fixing it, much like Evel

Knievel's long recoveries after his adventures on rocket-powered motorcycles in the American West. We drove a remote-controlled stuffed badger, RoboBadger, at the marmots and broadcast different alarm calls back at them. We found, somewhat unexpectedly, that none of the species produced predator-specific calls. Indeed, a human walking toward marmots could elicit many of the different call types that these species produced. So instead of producing predator-specific alarm calls, the marmots seemed to communicate degree of risk they experienced. And they did so in a remarkable variety of ways.

The golden marmots in Pakistan varied the number of calls they packaged together. These animals, living in a vibrant and intact predator community, would begin calling at me when I was over 100 meters away. As I approached, the number of calls packaged together into multinote utterances decreased. Thus, golden marmots tracked risk by varying the number of calls produced. As risk increased, they produced less conspicuous calls. By contrast, yellow-bellied marmots increased the rate of calling as risk increased. They did so with people but also called at higher initial rates when they detected an eagle or our model eagle.

As previously described, my more recent studies suggest that one of the functions of marmot calls is likely to communicate directly with the predator, perhaps to signal that it's been detected. Indeed, Erin Shelley, then an undergraduate honors student in my lab, and I conducted an evolutionary analysis of alarm signaling and concluded that the initial function of alarm signaling was likely directed to predators to discourage pursuit. We found that in 209 species of rodents, the evolutionary origin of alarm signaling was not associated with the evolutionary origin of sociality but rather with the evolutionary origin of being active during the day. Rodents active during the day seemingly benefited by emitting alarm calls directed to their predators. Further, while none of the species had referential abilities they had evolved a variety of ways to communicate risk. By packaging calls, varying the rate, and yes, by producing different call types, marmots communicated degree of risk to other marmots and potentially communicated to their predators that they had been detected.

Another interesting finding was the variation in call repertoire size. After recording hundreds of alarm calls in each species and making spectrograms (voice prints) that plot time and frequency (the pitch) and amplitude, I could sort calls by type. The critically endangered Vancouver Island marmot produced five different types of alarm calls in response to predators. By contrast, most other species of marmots produced one or two call types. Although Vancouver Island marmot alarm calls were not produced with sufficient production specificity to be considered referential, remarkably the order that we broadcast calls back to marmots influenced the magnitude of their response. It was as if Vancouver Island marmots had simple syntax! To note, only a few studies suggest that nonhumans have syntactical abilities—and syntax is one of the key traits that characterizes human language.

To expand our discussion of alarm calls, let's consider meerkats—small, adorable, and highly social carnivores. They live in the deserts of southern Africa where they forage for insects, cooperatively defend vulnerable young, and face existential threats from a variety of aerial and terrestrial predators. My friend and colleague Marta Manser, a professor at the University of Zurich, has studied meerkats for years in the Kalahari Desert of northern South Africa. She now directs the Kalahari Meerkat Project, the site of the popular television show *Meerkat Manor*. In this program cameras followed around Flower and her fellow soap opera stars through the trials and tribulations of life as a meerkat in the Kalahari. In the background, Marta was conducting elegant experiments to understand their communication system and the meaning of their vocalizations. Marta discovered that meerkats have referential signals—they produce different calls in response to ter-restrial and aerial predators. And importantly, they can simulta-neously communicate risk. Marta aimed to discover the types of in-formation they communicate in their calls. As she has shown, meerkat calls get noisier (remember, fear-inducing sounds are noisy) as risk in-creases. Thus, meerkats produce either a low-risk or high-risk call in response to an aerial predator. Meerkats that hear these different calls respond accordingly.

So what have we learned from this collection of studies? First, nature is grand in its diversity, and detailed study is needed to tease out the specifics. There are a number of putative adaptations to avoid predation, many that involve emitting alarm calls. Presumably, all work sufficiently as most species are not going extinct due to a faulty antipredator alarm system. Communication systems illustrate the diversity of adaptations that have evolved to solve similar problems.

While much of my comparative marmot work was designed to understand the rudimentary steps of the evolution of language-like communication, in retrospect this focus may blur the remarkable diversity of life that surrounds us. This diversity is ripe for study and as a source for creative inspiration. Biomimicry, a term coined by Janine Benyus, is a field that uses ideas from nature to solve human problems. For instance, engineers created body suits for Olympic-class swimmers after observing the sleek movement of sharks and analyzing their smooth skin. Geckos are renowned for their ability to climb vertical glass surfaces, and the structure of their footpads has inspired new adhesives. Much of biomimicry seeks novel solutions from physics and chemistry.

What about biomimetic lessons from the diversity of antipredator behavior? Can our knowledge of behavior be applied to increase our human security and defense? One such lesson is almost trivial—don't overreact. All successful animals have evolved from ancestors that were able to manage their risks well. And we know that across the scale of biological organization—from immunological responses to disease threats to behavioral responses upon detecting predators—it's extremely costly to overreact to a threat.

Physiological reactions include autoimmune diseases, which may be a fatal overreaction of the immune system to noninfectious stimuli. The immune system has evolved to respond to infectious challenges. Normally, it tolerates commonly encountered stimuli. But when your immune system is primed, these harmless stimuli are suddenly viewed as threats. Energy is misallocated to unnecessary defense, and the body itself is attacked. Those who study Evolutionary Medicine are exploring novel therapies to counter autoimmune diseases. And some of these therapies may capitalize on providing immune systems with benign threats, like nonlethal parasites, and thus, in a sense, occupy or otherwise distract the immune system from attacking the body itself.

On a much larger scale, consider the US response to 9-11. Over 3,000 people were killed, more than 6,000 injured, and countless more lives were ruined in the aftermath of these terrorist attacks. But according to the US National Highway Traffic Safety Administration, 10,874 people were killed from drunk-driving related crashes in 2017 alone. We will discuss in Chapter 12 some reasons why we react in horror to terrorist attacks in New York but not the constant and widespread toll exacted by drunk driving. If we had the mental clarity of *Star Trek*'s Mr. Spock we might be puzzled by the American reaction to 9-11. Spock may have suggested we further reduce legal blood alcohol levels for drivers, impose stiffer penalties for drunk or alcohol-impaired drivers, and better enforce seatbelt laws. Rather, in our quest for security, we have created what has become an endless war in Afghanistan, resulting in substantially more pain and suffering. By early 2020, there had been over 2,300 US military deaths and over 20,300 US military wounded. This doesn't include deaths and injuries of US government contractors, American allies, or the long-suffering Afghan people, nor the horrific effects of PTSD suffered by returning service members or by Afghans. Wars, once started, are often difficult to end. Overreaction to a threat can be costly, leading to death and impoverishment.

We should not forget that a marmot which overreacts to an alarm call by never leaving its burrow will eventually starve, and a marmot that forages foolishly may be killed. These trade-offs became very clear in March and April 2020, as the world reacted to the COVID-19 pandemic by shutting down most movement. If we listen to our inner marmot, we will know that individuals and nations that neither cut themselves off from others nor overinvest in defense will ultimately outcompete those that do. Successful marmots keep in touch and work together, something that was tragically lost in the initial nationalistic responses to the coronavirus pandemic.

Another lesson from risk communication is that alarm calls are often individually distinctive. Why would such an individual signature matter, and what purpose would it serve? Of course it could be an unselected by-product of how sounds are produced. In mammals and birds, air is forced through a vibrating sound production organ (the larynx or syrinx), and the sounds resulting from these vibrations are filtered by the vocal track to create the sounds that we ultimately hear. Morphological variation in the vocal tract could be responsible for idiosyncratic differences in vocalizations. Work with Kim Pollard, then a PhD student, revealed that in marmots, ground squirrels, and prairie dogs, species that live in typically larger social groups have more individually distinctive vocalizations. And I found that the acoustic features that seemingly communicate individual differences among marmots do not degrade as much when they are broadcast through the environment as those features that communicate risk. Both lines of evidence suggest that individuality in alarm calls has been subjected to natural selection and is not simply the unselected by-product of morphological variation in the vocal tract. In other words, a distinctive vocalization serves a purpose. We can infer past selection when we see variation in some trait—in this case, individuality—map nicely onto environmental or social variation—in this case, group size. Indeed, it makes sense that contact calls—vocalizations given by some primates, meerkats, and social birds that are used to keep track of others of the same species (conspecifics)—will be individualistic because you can't keep track of different individuals if they all sound the same. If there is a need to sound different, we expect there may be selection on the signaler to produce individually distinctive contact calls and selection on the receiver to distinguish them. One great example of the necessity of distinctive contact calls can be found in crèching penguins and marine mammals, who must find their young among a noisy and smelly colony with hundreds to thousands of other screaming young so that they can feed them when they return from foraging sorties.

Both adults and chicks produce and respond to individually distinctive vocalizations.

But how does all of this relate to our understanding of alarm calls? We know that marmots and other ground squirrels have individually distinctive calls, but what's the benefit to either the signaler or the receiver in producing or distinguishing them?

Reliability assessment may be the key. In Chapter 7 we learned that all individuals may not be reliable. Recall Aesop's fable about the shepherd boy who cried wolf. It is a clear representation that individuals vary in reliability. Moreover, we have a mechanism to explain how this may come about. We know that stress hormone levels can influence the probability of emitting calls, thus individuals could have different calling thresholds, which are modulated by stress hormone levels. If so, then some individuals may be generally more likely to call in response to both real predators and nonpredators, while others may call only when there are predators around. Let's call these individuals Nervous Nelly and Cool-Hand Lucy. Cool-Hand Lucy is a reliable caller, whereas Nervous Nelly is unreliable. If Aesop's fable is accurate, then we should expect Nervous Nelly to be ignored, much as the boy who cried wolf was eventually ignored.

We asked whether this hypothesis could explain why and how marmots respond to alarm calls. Because we were unable to properly estimate reliability, as it's really hard to know what stimulus triggers alarm calls in many situations, we conducted a playback experiment. We created a reliable marmot by pairing her calls with the presence of a stuffed badger and an unreliable marmot by pairing her calls with the presence of no badger. We employed a nifty experimental approach that Dorothy Cheney and Robert Seyfarth originally borrowed from developmental psychologists who study preverbal children. The technique is called a habituation-recovery protocol. This is the same technique used in Chapter 2 to study spider recognition in human infants. Again, when properly used, the technique permits us to make inferences about how animals classify different stimuli. In this case we first asked how an individual marmot responded to a random set of calls from a novel marmot. We did this for a number of different individual marmots, and we called this the pretest stage of the experiment. We then created a set of reliable (R) and unreliable (U) callers by broadcasting their calls with or without the badger present to the same set of individuals we had just pretested.

We then tested to see if marmots could distinguish callers based solely on their reliability. We broadcast either a new call from R or a new call from U. If R and U represented reliable and unreliable callers, we would expect two things if marmots had learned about caller reliability. First, they would increase their responsiveness to the reliable caller. Second, they would decrease their responsiveness to the unreliable caller. In the case of our playback experiments, we baited animals to a central location, broadcast the sound, watched them respond by looking, and quantified the time it took them to resume foraging. If the boy-who-cried-wolf hypothesis about reliability was to be supported, we expected to see marmots who heard the unreliable caller resume foraging sooner following the playback.

What we found was the exact opposite of what we expected: marmots hearing the unreliable caller stopped foraging for a longer period of time, mostly using this time to look around, while those that heard the reliable caller initially looked around and then resumed foraging. After dismissing the results as a silly mistake, we thought deeply about the reasons why marmots might respond more to unreliable callers. After all, these results are inconsistent with the boy-who-cried-wolf hypothesis, which has been suggested to explain individual specificity in the alarm calls of so many animals, including steppe marmots, Richardson's ground squirrels, bonnet macaques, and rhesus macaques.

What, after all, does reliability mean? It means that you can reliably infer something. By contrast, unreliability means that it will be difficult to reliably infer something. If what you're inferring is the presence of a predator, then perhaps it does make sense that after hearing an unreliable individual, marmots looked around more to independently assess what the true risk of predation was. It's as if they knew that even reliable animals make the occasional mistake, but you really can't trust the unreliable ones. Thus, unreliable callers or situations induce independent investigation.

Other lines of evidence support this idea. Marmots pay more attention or are more responsive when they encounter uncertain situations. They forage more after hearing alarm calls from older and presumably more reliable marmots compared to potentially unreliable calls from pups. Marmots also forage more after hearing undegraded calls compared to acoustically degraded calls. All sounds degrade as they are transmitted through space, and the farther they are transmitted, the more they degrade. The idea, therefore, is that undegraded calls simulate a situation where the caller is nearby; the sound was transmitted a short distance and was minimally degraded. Thus, after an immediate look up and around for the source of the alarm call, marmots resume foraging. When they hear degraded calls, they look around more, presumably because they are less certain about the true risk of predation. A degraded call could mean a caller is looking at you or away from you—and one looking at you might mean that the predator is nearby! So the take-home message is that unreliable individuals or situations are unreliable specifically because it's difficult to assess the true risk of predation. Given this difficulty, risky situations elicit independent investigation. And as with the production of alarm calls, we should generally expect evolutionary flexibility in mechanisms of communication.

Reliability assessment is likely to be a general explanation for the evolution of the ability to discriminate among callers. This illustrates another possible biomimetic response—when a situation is truly uncertain, some animals allocate more effort to assessing it correctly. This may (or should) sound a little familiar. Let's think about how we humans respond to uncertain sources of information. We are confronted with a twenty-four-hour news and spam cycle that churns out vast amounts of potentially contradictory information, including some that is entirely erroneous. Assessing the risk and reliability of this information is now more important than ever. We all have an inner marmot; we have evolved mechanisms to believe trusted sources. But our evolved systems can't keep pace; we face an evolutionary mismatch in which our evolved mechanisms have broken down, and we are apt to believe untruths or exaggerations. There's simply too much potential information to process. Mindful of this, I suggest that we scrutinize our news sources. If it sounds too good to be true, it could be a scam. News sources that follow strict journalistic practices and fact check their sources are, without question, going to be more reliable on average than those that simply aggregate information. The rise of fake news means that we must relearn how to trust but verify. And we must support reliable journalism that properly fact-checks sources because there simply isn't enough time for each of us to fact check everything we hear. We need good information to make informed decisions.

Animals use a variety of ways to communicate alarm signals. Anything that is produced in response to a threatening situation has potential signal value for an information-seeking recipient. Some rodents, such as kangaroo rats, rapidly beat their hind feet against the ground in elaborate bouts of foot drumming. Some of this foot drumming has been shown to signal territory ownership, but species also foot drum in response to detecting predators or cues from predators. Indeed, snakes have rather poor hearing but feel vibrations quite well. They are often the target of foot drumming by rodents keen to move them out of their territories. Conspecifics hearing a bout of foot drumming may wisely increase their vigilance and look around for snakes.

Besides foot drumming there are a variety of other sounds associated with increased risk. Some pigeons and doves, when alarmed, fly up and away and produce whistles with their wings. Rob Magrath's group and my students and I have both studied the signal value of these whistles. Rob and his students worked with the crested pigeon, an Australian pigeon with a distinctive spiny crest of feathers coming off its head. They discovered that when pigeons feeding on the ground were suddenly startled, they took off at a steeper angle than they would have if they were not as alarmed. These pigeons' specially modified feathers produced a whistle when they took off at a steep angle—an honest indicator of perceived threat. When the researchers broadcast alarmed wing whistles versus other flight sounds, they found that wing whistles were much more likely to cause a flock to take flight than non-alarmed wing sounds. Thus, wing whistles have the ability to communicate risk nonvocally.

In the course of research in the Virgin Islands, my students and I studied zenaida doves, the national bird of the nearby Caribbean territory of Anguilla. Like crested pigeons, zenaida doves produce mechanical wing whistles when they take off quickly. We wished to know how doves responded to them compared to other possible sources of risk information. We conducted a playback experiment and broadcast wing flaps to doves that contained whistles and wing flaps that did not contain whistles. We found that doves increased vigilance significantly more in response to wing whistles than to wing flaps without whistles (or in response to control playbacks). These results indicate that conspecifics interpret wing whistles as alarm signals. We then conducted another playback experiment which demonstrated that playbacks from a potential predator, the red-tailed hawk, elicited higher levels of dove vigilance than the wing whistles of fellow zenaida doves. Taken together, these results suggest that, contrary to other species, zenaida doves seemingly consider predator vocalizations more informative than conspecific alarm signals. This means that the quest for reliable information should be broad and not restricted to a single species or source of information.

As we have seen, wise prey use any cues they can to accurately assess predation risk, including the alarm calls that other species produce or the cues produced by predators themselves. What's particularly interesting about communication between species is that different species face different risks. Calls from another species may provide less reliable information about the true risk of predation. A number of playback studies have shown that mammals are likely to respond to alarm calls of other mammals as well as alarm calls from birds. What general factors might influence whether or not one responds to another species?

As a rough approximation, the size of an individual or species explains much about what it should fear. Small fishes are incredibly vulnerable to predation, and as they grow, they have fewer predators. Small mammals also may be vulnerable to a larger suite of predators. If so, a larger body size may be protective. As an aside, this likely explains why our corgi, Theo, was fear conditioned by the direct hit of nearby lightning—the extremely loud clap of low-frequency thunder could be interpreted as a nearby HUGE animal. Thus, vulnerability is related, at some level, to body size, and we see a variety of adaptations at both the individual and species level to counter these risks.

If the enemy of my enemy is my friend, how does body size influence the value of information? Imagine a small golden-mantled ground squirrel hearing an alarm call from a larger-sized marmot. What information could the squirrel infer from this? Or what information about predation risk could a marmot obtain from a hearing a squirrel? It turns out that wise squirrels should listen to marmots, but wise marmots may sometimes ignore squirrels. This is because everything that eats a marmot will eat a squirrel, but the opposite is not necessarily true. Marmots are about seventeen times larger than these squirrels. If this size discrepancy alone was responsible for explaining response, then mule deer, which are about sixteen times larger than yellow-bellied marmots, should ignore marmot alarm calls. It turns out that mule deer hearing marmot alarm calls immediately looked toward the speaker and sometimes fled. But why should deer listen to marmots when marmots don't listen to squirrels?

The answer may lie in sharing important common predators. At our Colorado study site both deer and marmots are preyed upon by coyotes and, before wolves were hunted to extinction, wolves. Thus, sharing important predators may trump size discrepancies. Information is potentially valuable even if it comes from a much smaller animal. This lesson is highlighted by the fact that vervet monkeys respond to alarm calls by the diminutive superb starling, a brilliantly colored bird found in East Africa. Even though vervet monkeys are sixty-three times as large as starlings, they share some subset of both aerial and terrestrial predators. Vervets are able to respond to starling alarm calls, an ability that is certainly the result of learning since the species do not coexist throughout their ranges. However, they are quite discriminating and can be trained to ignore all calls from unreliable starlings.

Much as a peacock's call can be used to find a tiger, all organisms naturally seek information about risk from a variety of sources. But all sources are not equally reliable. It's essential to evaluate a source's likely accuracy before acting on it. Acquiring information about risk is essential, even if it comes from other species. Alliances, whether evolved or learned, should be expected. Thus, maintaining intact communities of animals may be vital for ecosystem stability and species persistence. We'll learn much more about the importance of fear in ecosystems in the next chapter. Regardless, a lesson for humans is that those tasked with assessing the risk of fearful events will benefit from seeking information from all who share similar risks.

9

CASCADING EFFECTS

In 1962 my mentor and colleague Ken Armitage began studying the marmots at the Rocky Mountain Biological Laboratory (RMBL). It is now one of the world's longest-running studies where researchers have followed the fate of individually marked nongame mammals, and it is a priceless opportunity to see how environmental and social traits influence population dynamics, including the costs and benefits associated with living in groups.

When I took over the day-to-day management of the RMBL marmot project in 2001, my immediate goal was to better document the duration of the marmots' four- to five-month active season and their behavior immediately after they emerged from seven- to eight-month-long hibernations. But I also realized that marmots appeared superbly adapted to dealing with predators. I aimed to drill down and learn as much as I could about their antipredator behavior.

Some of my happiest times in the field have been in April and early May, waiting for the yellow-bellied marmots to emerge from hibernation. I feel privileged to be able to ski to work in such a spectacularly beautiful location. In the winter and spring, the upper East River valley has no motor vehicles and no snow machines. While a few caretakers spend the winters keeping an eye on the Rocky Mountain Biological Laboratory, the area is mostly uninhabited and quiet despite being only eight miles up the hill from Crested Butte. Then, in April, Team Marmot arrives.

A routine day begins around 6 AM with a morning ski to reach the marmot colony sites. Then my assistants and I wait. Each member of our team typically works alone. This way we can visit a large number of colony sites each morning. By lunch the snow is often too soft to traverse and sticks to the bottoms of our skis and snowshoes. It is quiet and often cold on these April mornings. After the extreme exertion of skiing at 10,000 feet, we pile on clothes and sit for a few hours, scanning the snow above the burrows where we suspect marmots will emerge. Sometimes we're rewarded: a marmot pokes its head out of its snow-covered burrow for the first time that season. After burrowing up through two meters of snow, the sodden rodent emerges into the blinding light and shakes off a winter's accumulation of fleas. At times we have company in the cold. Like us, coyotes and foxes sometimes wait patiently near the marmot burrows. We do, however, have different goals in mind. While we sit hoping for a marmot sighting, the coyotes and foxes await a meal. I've seen a fox travel from marmot burrow to marmot burrow, scent marking each one as if to tell his peers, "these are mine!" But it's the coyotes that are the main spring predators.

In the spring we regularly see coyotes in the valley, shaggy and resplendent in their luxurious winter coats. Howling to the wind, these song dogs call to their mates and announce territory ownership. The coyotes frequently hunt the newly emerged marmots. Sometimes we see the coyotes course in, as if from nowhere, and grab an unsuspecting marmot. While I root for the marmots, I am delighted every time I get to watch an undisturbed coyote; later in the season the coyotes disappear into the shadows. As the snow melts and the valley road opens up, people arrive—tourists, researchers, and residents—and the coyotes shun us. On occasion we see them around a marmot colony, but we no longer see them around the Rocky Mountain Biological Laboratory. They seem to be afraid of us. Rather than seeing them howl during the day, we hear their howls mostly at night, and we see very few during the day. This fear of humans has ecological consequences on the valley.

In this chapter we will learn about these ecological consequences. They are common and occur not only in our Colorado study site but in many habitats. We will spend a bit of time in the Greater Yellowstone Ecosystem (in and around Yellowstone National Park) and then head back to Vancouver Island. We will also travel to the Australian outback for more examples. We will learn about some of the controversies surrounding what are called trophic cascades. We will also learn about the ecological and behavioral consequences of the loss of predators and the wildlife conservation challenges they create. Finally, we'll shift our focus to humans. Our perceptions of risk have important influences on where we live and may also influence community cohesion.

In spring, migratory mule deer leave their wintering grounds at the Almont Triangle State Wildlife Area in Colorado and travel to the subalpine valleys to forage on the verdant summer vegetation and have their young. By the last week of May they've made it up to Gothic, Colorado, the location of my long-term marmot study. Colleagues and I have found that females prefer to have their young at the Rocky Mountain Biological Laboratory town site in Gothic, an old ghost town that is now one of the world's premier high-elevation field-research stations. It is there that an international community of biologists study vertebrates, plants, insects, and microorganisms in the valley and produce long-term data sets. Pets are banned from laboratory property, and visitors and researchers alike try not to damage the resources that are under detailed study.

Together with my friends and colleagues Nick Waser and Mary Price, broadly trained ecologists who study plants and pollination biology at the lab, I researched the effects of the deer on the vegetation in the area. To do this work we pooled archival data and conducted new experiments on the deer's preferred plants. We found that the deer's preferred plants in the town site had reduced reproductive success because of the increased deer population. By positioning coyote urine next to salt blocks that attracted deer, we also discovered that the does avoided foraging around coyote urine, suggesting that it was the fear of predators that drove them to the safety of the town site. The lack of predators was ultimately responsible for different impacts on vegetation by deer, both in and outside the town site.

Predators have profound influences not only on the distribution and abundance of their prey, but also on what their prey eat, and potentially on the abundance of other predators and what *they* eat. In Chapter 4 we saw how smaller predators fear larger predators and are responsive to their scents. Ecologists would consider these behaviors direct effects of predation. But predation has indirect effects as well—effects of a predator on the distribution or abundance of another species that works through an intermediate species. For instance, more predators (such as lynx) may mean fewer herbivores (such as arctic hares). The result is in an increase of the plants that the herbivores would normally eat. This specific type of indirect effect, a "trophic cascade," illustrates how changes in what one species eats has cascading effects through the ecosystem.

Nick and Mary and I detected a trophic cascade acting through coyotes and deer at our Colorado field station. People's presence scared away the coyotes, which made it a relatively predator-free place for deer to graze and raise their fawns. The absence of deer led to declines in the vegetation only in the town site, where they preferred to browse. Thus, certain plants (those preferred by deer) were indirectly influenced by the coyotes.

Darwin may have been the first to formally recognize trophic cascades. He wrote about the effects of house cats eating mice that, unchecked by the cats, would eat honeycombs, and therefore influence plant–pollinator interactions, resulting in changes in plant abundance. If you were to run a thought experiment and remove all the cats, the mouse population would increase and the plant population would decrease. Thus, the relative abundance between predators and prey have effects that extend to other species. In this case, the prey's prey—the bee larvae—also affect the pollination rate, and hence the success of plants.

Trophic cascades can be driven by direct consumption, such as when lynx eat hares, or they can be driven by the fear of predation that influences where prey eat. Lynx cues scare hares into avoiding certain areas, for example, and the vegetation that hares prefer increases in those now hare-free areas. As we have seen before, prey are quite responsive to the many predatory cues, and upon detecting these cues, they may change their activity patterns and avoid particularly risky areas. As we saw in Chapter 3, when Liana Zanette and Michael Clinchy broadcast predator vocalizations, these sounds alone were sufficient to reduce song sparrow nesting success.

Following a seventy-year absence, wolves were reintroduced into Yellowstone National Park in late 1995. By 2016, the wolves had experienced such rapid population growth and spread that they were removed from federal protection. This result initiated an ecosystem-wide experiment into the ecological role of top predators and stimulated considerable debate among federal employees, conservation biologists, ecologists, and ranchers. While some biologists say that the wolves' reintroduction to the Yellowstone Ecosystem set the textbook example of a trophic cascade in motion, there is some controversy over whether this is driven by predation alone, by fear alone, or by a mix of these factors.

The distinction between fear and predation and the identity of the one that drives the cascade is important because the wolves live in areas where ranchers run livestock, and there is big-game hunting of elk, moose, and deer. Clearly predators prosper by eating prey, but are the broader ecological effects on vegetation driven by the presence of fewer herbivores because the predators have eaten them, or are these effects driven by the fear that the predators create, changing the distribution and reproductive success of their prey? Fear alone can reduce population size and modify behavior. But hunters also change the dynamic by hunting and eating wild animals, and ranchers find their livestock are killed by wild carnivores. If the wolves have a punishing effect on livestock and elk populations, critics point out it should be acknowledged.

I heard the debate firsthand at a conference hosted by the Yale Institute for Biospheric Studies. I was invited to speak there as the "prey guy" in the midst of discussions on the effects of large carnivores in terrestrial ecosystems. To my benefit, over the conference's two-and-a-half days I learned about the ecological, social, and political effects of carnivores throughout the world from an international Who's Who of carnivore experts.

Dan Ashe, then the director of the US Fish and Wildlife Service, spoke first, recounting the lessons he learned from the wolf reintroduction. Dan's talk was one of the most honest and self-critical talks by the head of a government agency that I have ever heard. He said that if he were to do it again, he'd worry less about the wolves (which did fine because populations increased quickly) and much more about the relationships with ranchers and the community (which remain strained). Coexistence with carnivores in human-dominated landscapes is a political and social challenge, driven, in part, by our fear and loathing of carnivores. Fear, after all, does have consequences!

The debate about whether trophic cascades can be driven by fear began on the first full day. Doug Smith, the longtime director of the Yellowstone wolf project, was the first speaker. Doug summarized the challenges of restoring large carnivores in places where humans live and work and some of the successes. Matt Kauffman, a researcher with the US Geological Survey and a professor at the University of Wyoming, was a panel member in the discussion that followed. He criticized the evidence for the fear-mediated cascade. His research led him to believe the mechanism (fear or predation itself) really matters—because fear somehow implies that carnivores are not killing animals, which, after all, they must do to survive.

One of the first comprehensive studies and reviews of all of the available evidence that fear alone can influence community structure was conducted by William Ripple and Robert Beschta at Oregon State University. An expert on the ecological effects of carnivores, Bill also is a leader in efforts to conserve carnivores and other megafauna, species with outsized effects on the ecology of their landscapes throughout the world. Bill and Robert began their review noting how wolves were first historically present, then absent for about seventy years, and then present again with reintroduction in Yellowstone. This large natural experiment helped define the collapse and restoration of a trophic cascade.

The trophic cascade they focused on was the effect of wolves on elk and the resulting effects of elk on soils and woody plants such as willows and aspen—highly preferred elk winter forage. They summarized a series of studies which showed that when wolves were absent, willows and aspen were unable to grow unless they were in fenced enclosures. They also summarized tree-ring data on aspen and cottonwood trees. By extracting a narrow core of a tree and counting annual growth rings, it's possible to age trees and study their condition. Thick annual growth rings mean that it was a good year; narrow annual growth rings mean that the year was not as successful. They noted that when wolves were present, the number of aspen and cottonwood trees declined.

Fifteen years after the wolf reintroduction, with established packs of wolves in Yellowstone, they collected new data on aspen and cottonwood and reviewed available evidence about the relationships among wolves, elk, and woody plants. They found that while all aspen were foraged on by ungulates soon after wolf reintroduction, browsing intensity declined, precipitously in some places, years after wolf reintroduction. As browsing on young aspen declined, the aspen grew taller. Bill and Robert also noted that following the introduction of wolves, beaver populations increased, willow populations increased, browsing in riparian areas—close to rivers and streams—decreased, and elk populations decreased. They summarized a variety of studies in the Yellowstone area that have shown how elk behavior changed after the introduction of wolves. Elk modified habitat use to avoid riparian areas, changed their movement patterns, increased their group sizes to reduce predation risk, and increased their antipredator vigilance.

Bill and Robert concluded that these results illustrate a classic trophic cascade that linked more wolves to fewer elk, as well as modified elk behavior, along with the subsequent increase in woody plants, elk food. They noted that initially the effect of fear on elk behavior might be more pronounced and only later would be replaced by the effects of having fewer elk.

Finally, Bill and Robert noted in their review that songbird diversity and abundance had started to increase with the growth of riparian willows. They noted that the restoration of streamside vegetation stabilized the stream banks, was tracked by beavers, and reduced soil runoff, which created physical effects on stream quality. They summarized with the following statement: "Predation and predation risk associated with large predators appear to represent powerful ecological forces capable of affecting the interactions of numerous animals and plants, as well as the structure and function of ecosystems."

This research and analysis appeared pretty balanced to me, so why did Matt Kauffman believe so strongly otherwise? Matt, like Joel Berger, has spent many winters studying elk behavior and ecology in the subzero temperatures of northern Wyoming. To study the strength of fear effects, Matt and colleagues systematically surveyed aspens, adopting similar techniques used by Bill and Robert. Matt also considered other hypotheses that might explain changes in aspen growth and distribution.

Matt and his colleagues first questioned, on logical grounds, whether wolves, which hunt over large landscapes, can exert the sort of consistent fear that would have strong influences on prey, such as elk. This questioning is important because some reviews and studies have shown that the strongest behavioral effects are seen when predators are very localized because prey are making behavioral decisions about where to forage on the scale of meters. They emphasized that in terrestrial cascade studies wolves travel farther on their landscape than seen in many other systems. They also noted that aspen have declined throughout the US Intermountain West over the past few decades, and that elk are partially implicated in this widespread decline. The populations of elk and wolves were quite low before the creation of the national park in 1872. In fact, the managers were quite concerned about the damage elk caused on vegetation and instituted an annual elk cull that continued up to 1969.

The study by Matt and his colleagues used a decade's worth of data of the locations where wolves killed deer to precisely map what is referred to as the landscape of fear. And, importantly, they focused on a different part of the park (the Northern Range) from where Bill and Robert worked. Based on their detailed map of where elk were killed by wolves, they estimated the greatest predation risk for elk was in riparian areas.

Unlike many of the studies of large carnivore–driven cascades, which are based on "natural" experiments and are correlative, meaning that they look for patterns on the landscape, Matt and his colleagues also conducted an experiment. They established a series of fences to protect aspen from elk grazing and found that without this protection, small aspen stems were unable to grow, whereas in the protected areas, small aspen grew into larger aspen. This experiment eliminated alternative explanations for observed patterns. Their key experimental finding was that there was no relationship between where elk were likely to be killed and elk grazing pressure on young, unprotected aspen stands. This finding suggested that elk were not sensitive to variation in risk, and small aspen stems were not more likely to survive and grow in areas with high predation risk. Their comprehensive study was carefully executed and well justified. His team's experimental results, though accounting for a shorter time period and on a smaller scale, went against the correlative evidence that fear drove the dynamics of aspen and cottonwood.

Matt and his colleagues justified the decisions that they made, which differed in some of the details from Bill's earlier work. I believe we must accept their results as an accurate summary of what happened over the course of a decade at the locations they studied: elk modified their behavior in response to wolves and drove the decline in aspen, and the reintroduction of wolves did not rescue aspen. Indeed, at their study site they found no evidence of a trophic cascade. More importantly, at their study site they found no evidence of fear influencing elk's effects on aspen.

More recent work by Matt and collaborators has comprehensively found that elk are quite sophisticated in their habitat selection when living near wolves. Wolves hunt at predictable times of the day, and elk avoid the areas where wolves are when they are hunting. This creates a highly dynamic landscape of fear. The authors suggest that elk use risky habitats at safe times of the day. And this, they argue, is key to understanding why in their studies they did not find evidence of a behaviorally mediated trophic cascade.

So, what to make of these differing studies and the overall debate? We know that trophic cascades are common in both terrestrial and aquatic systems, but the strength of predators driving them may vary over both time and space. We know that fear can influence the reproductive success of a species, as Liana Zanette and Michael Clinchy's work on birds has so nicely shown. And we know that predator density has profound effects on the number of their prey.

I think that Bill and Robert's study, combined with the results of the study by Matt and his colleagues, shows that there is spatial and temporal variability present in the Yellowstone Ecosystem. The great wolf reintroduction experiment has driven cascades in some parts of the ecosystem, but perhaps other factors are more important in other parts of the ecosystem. Fear may have an important role in certain places, and these recent studies have shown how sophisticated elk can be in avoiding risky areas. As Joel Berger's work showed, behavioral responses may occur on the spot and may compensate for increased predation pressure. I believe future studies are required to get a more precise estimate of the relative influence of fear on the distribution and number of ungulates. More data will likewise inform us about the relative influence of direct predation on ungulate distribution and abundance.

While the effect of fear may be contested in Yellowstone, it's not in the rocky intertidal off Vancouver Island, Canada. There is a rich literature of marine trophic cascades driven by consumption, and Michael Clinchy, Liana Zanette, and Justin Suraci, a graduate student at the time, identified a fear-based cascade that works across the aquatic and terrestrial environment.

When I visited the site of their bird experiment years ago, Michael and I were considering how to broadcast predatory sounds to raccoons. On these bay islands, raccoons forage on crabs. As one would imagine, in areas where raccoons are present, there are fewer crabs and intertidal fish. It's rather funny to watch full-grown raccoons, standing on their tippy toes, back arched, belly fur moist with the ocean, pulling out and consuming colorful crabs from the rocky intertidal zone at low tide. We broadcast a raccoon predator—we used dog barks—to the raccoons, and they paused and looked around. Bingo! Justin, Michael, Liana, and other colleagues capitalized on this result by hanging speakers from trees and broadcasting dog barks (a potential raccoon predator) along with control sounds (harbor seals and Steller sea lion vocalizations) for a month at a time. And since Michael, an extraordinary field technologist, was involved in this experiment they used time-lapse video cameras to follow the results.

They found that when the dogs barked, raccoons looked up, decreased foraging, or fled the intertidal area. When exposed to a month of these playbacks, the same responses were seen, which led to a reduction of time spent in the intertidal areas. After a month there were 97 percent more intertidal crabs, 81 percent more intertidal fish, and large increases in intertidal polychaete worms and subtidal rock crabs. Also, because crabs compete with a particular fish—a sculpin—increases in crab abundance were associated with a decrease in fish abundance. And because the intertidal crabs eat periwinkle snails, as crab populations increased, snail populations decreased. These results convincingly demonstrated that fear of large carnivores had effects on multiple trophic cascades, and demonstrated for the first time that the fear of terrestrial predators can drive the dynamics of an aquatic ecosystem.

Wars can also create trophic cascades. Not long after independence from Portugal in 1975, Mozambique descended into a bloody, fifteen-year-long civil war. In addition to an estimated 1 million humans killed during the war, wildlife in Mozambique were decimated. Large herbivores were shot for food, predators were killed, and there was considerable illegal hunting of large African game animals. The ecological impacts of this persist today. Founded in 1960, Gorongosa National Park was not immune from violence or illegal hunting, and it's estimated that 90 percent of its large animals disappeared. This included almost all of its large carnivores. In 2004, the American philanthropist Greg Carr first visited the park and saw an opportunity to rebuild it. Carr negotiated agreements with the government of Mozambique and, working with the local people, began protecting the park from further poaching as well as reintroducing locally extinct herbivores. Science played a vital role in restoring the park, and he consulted with a variety of leaders in the fields of ecology and conservation biology. As you may have realized, a situation with few predators but growing populations of prey may have effects on the landscape, and such systems provide opportunities not only to study trophic cascades but to tease apart direct and indirect impacts that may drive them.

Justine Atkins, then a Princeton University graduate student, her advisor, Robert Pringle, and a team of colleagues seized the opportunity. They focused on bushbuck—an antelope that prefers, in predator-rich areas, to live exclusively in the forest. However, in areas where they are not threatened by predators, bushbuck will venture outside the forest to seek higher-quality food in more open riverine habitats. Justine and her colleagues were able to experimentally change both the distribution and diet of bushbuck and by doing so identify a fear-driven trophic cascade. They broadcast the sounds of predators and put out feces and urine from large predators. Then they followed bushbuck with GPS collars that recorded each individual's precise location every fifteen minutes. Compared to their use of space when they experienced nonpredatory control stimuli, bushbuck avoided areas with predatory stimuli. Moreover, those in more open habitats (which incidentally contained higher-quality food) more often moved to areas with trees at times when predatory stimuli were present. These and other results suggest that the widespread elimination of predators during the civil war modified a trophic cascade and let some herbivores move into a formerly risky location to forage on vegetation to which they historically had limited access.

Trophic cascades are found in many different environments, and human interventions, whether planned or unplanned, help us identify them. My friend and colleague Mike Letnic, based at the University of New South Wales, is a remarkably creative biologist, an avid fisherman, naturalist, and a constant source of ideas. Doing work in the field with Mike has given me many new insights into the nature of fear. He has been studying tropic cascades in a remarkable experiment featuring the nearly 3,500-mile-long dingo fence that keeps dingoes out of parts of South Australia, essentially all of New South Wales, and the southern part of Queensland.

Dingoes are either a subspecies of wolf or, as Mike has argued, their own taxon descended from wolves. Regardless of their provenance, dingoes are relatively recent inhabitants of Australia, arriving between three and five thousand years ago, likely with humans, from New Guinea. Humans, on the other hand, have been in Australia for at least 50,000 years, and their ecological effects have been well studied. The arrival of dingoes to the Australian mainland was associated with the extinction of thylacines, a marsupial wolflike animal, and Tasmanian devils, which are now restricted to Tasmania, an Australian offshore island. The dingo fence (also known as the dog fence) was built in the late nineteenth century to keep dingoes out of the sheep grazing areas of southeastern Australia. I suppose one could say this fence has achieved its purpose—dingoes doesn't cross it. But at a mere 1.8 meters high, foxes, cats, rabbits, and kangaroos have no problem crossing the wire mesh fence.

While kangaroos and other marsupials are native to Australia, the introduction of European red foxes, cats, and rabbits in the nineteenth century has taken a devastating toll on Australian fauna. Australia has the world's worst record of recent mammalian extinctions: over twenty species of mammals have been driven extinct through predation by foxes and cats since European colonization. Dingoes are very good at killing foxes. Indeed, it's common for larger predators to eliminate their competition—something ecologists refer to as "intraguild predation." Such intraguild predation has left very few foxes on the northern, dingo-rich side of the fence.

Mike and I, along with Katherine Moseby, another Australian friend and conservation scientist extraordinaire, walked and compared tracks along both sides of the fence. There are essentially no tracks from small mammals on the side of the fence where dingoes are absent; the foxes and cats have eaten them all. Later, Mike showed me photographs from his studies taken on either side of the dingo fence at a number of different locations. These images showed the stark differences in vegetation. Where kangaroo- and rabbit-eating dingoes were present, there was more vegetative cover. Where dingoes were not present, there was noticeably less vegetation. Additional studies confirmed that mammal communities on paired plots on either side of the dingo fence also differed due to dingo predation.

Katherine and her husband, John Read, created an amazing experiment in the South Australian arid zone. On land owned by the world's largest uranium mine and the world's fourth-largest copper mine, Arid Recovery was born. The nonprofit they created (but no longer run) conserves and restores local biodiversity. Katherine and John established a 123-square-kilometer predator-proof fence. The reserve is divided into different fenced paddocks, most on the scale of twenty-five square kilometers. Once they perfected the fence design to ensure predators couldn't get in, they began introducing animals, including those that had been driven extinct on the Australian mainland.

Burrowing bettongs, cat-sized rat kangaroos, are the only macropodid marsupial to dig burrows. In the last century these highly social and endearingly cute animals were driven extinct by foxes and cats throughout the Australian mainland, yet three known populations persisted on small, mostly predator-free islands off the western Australian coast. These populations were the source of the Arid Recovery animals. The initial twenty-eight colonists grew rapidly without terrestrial predators into a large population of around 8,000. As expected, the bettongs, unchecked by predators, quickly consumed all of the vegetation. Katherine and John moved one group into other enclosures and another group outside the fenced enclosures. Once outside the predator-proof fence, the animals were quickly eliminated, likely by dingoes as well as cats.

Katherine focused her next experiment on greater bilbies, a creature resembling a piglet with huge linear ears, a Pinocchio-like snout, and a body shaped like a rugby ball. She used an aversive prerelease training similar to that used by Andrea Griffin with tammar wallabies (as discussed in Chapter 7). Katherine and colleagues trained bilbies to respond aversively to the sight of cats or cat olfactory secretions. A key metric she used was whether bilbies moved more or changed burrows after encountering predatory cues at one burrow. Bilbies showed both of these responses in the predator-free enclosure where they were trained. But she later found no differences in burrow use or movement following the release of ten trained and ten untrained bilbies outside the fence to an area with free-living cats and foxes. She concluded from this that training was ineffective in successfully modifying antipredator behavior.

Because of her failure to change behavior of animals trained in rather unnatural situations, Katherine contacted me, suggesting that we should instead let natural selection choose which animals could be released outside the fence. In other words, she contemplated letting the captive animals live with a few cats before reintroducing the survivors. I questioned the ethics of this, not realizing the vast scale of the enclosures at Arid Recovery. We discussed the possibilities for about a month, agreeing that we both learned a lot from the vibrant discourse. She asked if I wanted to join her and Mike on a grant proposal she was developing to test these ideas. I immediately jumped at the opportunity to learn with her.

Two years after the initial correspondence, we walked along the Arid Recovery fence line. Unlike the dingo fence, this fence is much taller and has a floppy top that prevents foxes and cats from success-fully climbing in. Conveniently, the dingo fence bisects Arid Recovery so we could see the ecological effects for ourselves. There were no small mammal (other than rabbit) footprints on the dingo-free, fox-rich side of the dingo fence, but inside the Arid Recovery fence, we found many footprints. North of the dingo fence, there were a few footprints, but again, inside the Arid Recovery fence, we observed many footprints. It appears that fencing is a highly effective way to grow bettongs, bil-bies, and other small mammals in Australia! However, fencing is not sustainable in the long run for the growing populations of bettongs and bilbies. As noted previously, bettongs overeat the natural resources inside the fence. Bilbies will do the same when their population in-creases. In this new set of experiments we aimed to see if exposing ani-mals to predators—cats in this case—was sufficient to prepare them for life outside the fence.

We called this process *in situ* predator training because we allowed learning (and possibly natural selection) to occur more or less in a natural environment. After five years of research we have discovered that bettongs and bilbies can thrive while living with low densities of cats, and their antipredator behavior increases when exposed to low densities of cats. Since cats were recorded killing only a few individ-uals, all of these behavioral changes are induced by the fear of cats. While we can't follow individuals and don't know their exact interac-tions, we can assume that bettongs and bilbies have had encounters with cats, and these experiences have changed how they behave around them. Specifically, we found that living with cats increases bettongs' ability to discriminate the smell and sight of predators from nonpreda-tors. In response, they became more wary. Thus, living in a fearful environment has likely made these species more prepared for life out-side the fence—a formal hypothesis that we're actively testing.

Could this kind of *in situ* exposure to perceived threats be helpful for humans? Based on what I have seen in my experiments, it may be worth considering. After all, we see the ecological consequences of fear in animals, and it isn't that much different from what we observe in our societies. The tendency to avoid certain neighborhoods perceived as dangerous results in a lowered population and reduces the productivity of those neighborhoods. But why do some neighborhoods generate fear? Is there a legitimate reason?

Current crime statistics may be one factor that drives fears, and this is seemingly a rational response. I acknowledge that in modern society the vast majority of crime does not, like predation of bushbuck, end in death. But the fear of violence is a healthy response, one that may preserve our resources and fitness if not our very lives. We've seen how we have a well-prepared suite of neurochemical responses that ensure that we live another day. These often prevent us from getting into fatal encounters. Thus, we're more like bushbuck than we might like to admit!

But sometimes our fears are simply of those unlike ourselves, whether these "others" are people of a different color or from different socioeconomic backgrounds. We are primed, as Robert Sapolsky writes in *Behave,* to create what he calls "Us versus Them" categorizations. Such categorizations may be based on nationality or sporting teams or colleges or fraternities or religion or gender or race. These categorizations are part of the human experience and lead to tribalism.

Can exposure to those we fear inoculate us against tribalism? Sociologists have studied factors influencing community cohesion for years. Robert Putnam's *Bowling Alone* is a must read. He describes how we've become isolated and fearful and discusses the negative consequences for community cohesion and societal function. There are also studies that focus specifically on ethnic diversity and perceived fears.

The effects of ethnic diversity on social cohesion are complex and seemingly depend on whether someone initially perceives other groups as threatening. For those who do, increased exposure to other groups may, unfortunately, drive them to be more mistrustful and therefore reduce social cohesion. One study suggested that this is enhanced for those in lower socioeconomic classes and speculated that it's not diversity, per se, but inequality that enhances perceived threats from others. More research is clearly needed to understand how to reduce tribalism and discrimination and how to build more cohesive and less fearful societies. Perhaps a foundational understanding of why we fear what we fear may produce novel insights.

Fear among middle-class and wealthy homeowners may drive lags between the beginning of planned gentrification and neighborhood transformation. Perhaps this is influenced not by current crime statistics but the memory of previous crime statistics—the ghosts of predation past. Wise investors who can separate the current risks from the historical risks will often see their investments skyrocket.

When Janice and I moved to Los Angeles, the memory of the LA riots was still fresh. We wanted to see houses in Culver City, then an affordable neighborhood with decent schools about five miles from UCLA. Jim, our real estate agent, cautioned us against Culver City and pointed out the area we wanted to explore was just three blocks away from a large shopping center with a Target store. "Remember the *New York Times* photo of the burning Target?" he asked.

That part of Culver City was then and is now a great family neighborhood. Everyone knows everyone, and the kids move in packs between houses—something less rare than you might think in Los Angeles. The area is now replete with highly paid people who work at Silicon Beach—a tech hub not far from this part of Culver City—and these tech workers have driven up the price of real estate. Had we bought a house there, our property value, less than twenty years later, would have increased by three or four times. We allowed past fears of a dangerous environment to cloud our perceptions of current risk.

Fear has a variety of cascading effects on communities, as we have seen in Yellowstone National Park, on small islands near Vancouver Island, in Mozambique, and in the Australian outback. Despite the fact that these effects are often caused by predation, fear of predators itself may have a variety of consequences that range from prey changing their activity patterns (as we have seen with Yellowstone elk), investing less in their offspring (as we have seen with song sparrows), and influencing decisions about where to forage (as we have seen with raccoons).

Our fears, whether real or perceived, influence where we live and perhaps what we do. Our disproportionate fear and loathing of large carnivores, such as wolves, threaten them with extinction and change the ecosystems that they regulate. Yet each year *many* more people are killed in car accidents than by all of the carnivores on Earth. Why is it such a challenge to live with large predators, yet we readily accept the risks of car accidents? In the next chapter we will learn why we may overreact to certain threats and underreact to others.

10

MINIMIZING COSTS

I first met Rafe Sagarin, a marine ecologist and visionary thinker, after a talk I gave on predation. Along with his work and teaching at the UCLA Institute of the Environment and Sustainability, Rafe organized a National Center for Ecological Analysis and Synthesis (NCEAS) working group on the subject of "Darwinian Homeland Security." NCEAS was a premier government-funded think tank based in Santa Barbara with an aim to increase ecological knowledge. After 9-11, when Rafe was a Congressional Science Fellow in Washington, DC, he realized that the security at the Capitol and around town was predictable and, therefore, easy to penetrate. He wondered if lessons from four billion years of life could offer insight to improve our security systems. He wrote an article in the journal *Foreign Policy* that laid out the basic logic of his ideas and what we later came to define as the field of natural security.

At the first meeting we introduced ourselves. Group members included a paleontologist who has been the recipient of a MacArthur Fellowship, a co-founder of the field of evolutionary psychology, a colleague with doctorates in zoology and political science, an expert on suicide terrorism, and a military officer turned bioweapons inspector turned peacemaker for a nongovernmental organization. I eagerly looked forward to learning as much as I could from this diverse group. Rafe had just read a book about running meetings with no preset agenda and allowing attendees to create their own together. So we started to talk broadly about what we wanted to discuss over the next several days.

Through our collaborative discussions we created the highly integrative field of natural security. Our efforts were later captured in the book *Natural Security*, edited by Rafe Sagarin and Terry Taylor, the former bioweapons inspector in Iraq before 2003. Many of the meeting participants contributed chapters. Later, Rafe wrote a book, *Learning from the Octopus*, which effectively captured some of the excitement in the room during our many discussions and organized them into a checklist for creating resilient and adaptable systems and enterprises.

Rafe later moved to Tucson, Arizona, and built a coral reef in the middle of the desert, leading a project at Biosphere 2. He continued to lead our expanded natural security group, and we continued to collaborate. In late May 2015 at 4:50 PM, Rafe emailed me his penultimate version of a paper we were working on together. Then he hopped on his bike for a ride. He was tragically killed by a drunk driver about twenty minutes later. The loss is indescribable. He was a brilliantly synthetic and integrative thinker and an outstanding scientific communicator. He had many more great ideas and projects left to realize. But his leadership lives on through the field of natural security and the connections he made for and with people.

At the NCEAS meetings I was introduced to Dominic Johnson. Dominic is the aforementioned political scientist with double degrees who eagerly absorbed lessons from human and animal behavior, neuroscience, and decision theory. Dominic's knowledge is voluminous; he's comfortable talking about wars through the ages, territoriality in European badgers, positive illusions, or why humans have repeatedly invented religion. This wonderful fluency in different fields helps him create genuinely novel insights, in part about how we make decisions. We'll discuss the importance of his theories about human overconfidence and war slightly later in this chapter. While on sabbatical at UCLA, Dominic had an office near mine. Over coffee we would fill our respective white boards with an outline that drew together links between theoretical and empirical results from very disparate fields. Error management theory—an idea that uses data to help us understand the conditions under which we should behave one way or the other—emphasizes the importance of minimizing long-term costs. In the process of sharing our varied data, Dominic and I found to our surprise that optimal decision making often included a slightly biased estimate about the true nature of the world.

Martie Haselton, a colleague at UCLA, is an authority on error management theory, which is also known as signal detection theory. Signal detection theory assumes there are two ways to make a detection mistake. First, you can miss detecting something risky, for instance an eagle, that is present. This is called a false negative mistake. Second, you can respond fearfully to something that is not risky, say, a vulture that superficially resembles an eagle. By doing so, you have made a false positive mistake. Importantly, as you reduce the chances of making one type of mistake, you increase the likelihood of making the other type of mistake. This is because detection decisions involve setting what's called a detection threshold. If you're too selective, you'll never be wrong; you'll not make any false positives and will never assume that a vulture is an eagle, but you will inevitably miss things that are there and thus make false negative mistakes (you will miss some eagles). Fundamentally, signal detection theory emphasizes the inevitable trade-off between false positive and false negative mistakes.

Error management theory permits us to understand decision making under uncertainty. (Was that quickly moving object a predator? Should I stop foraging because I've smelled a predator that could be here or not?) Since virtually all decisions are made under some uncertainty, error management theory helps us understand why we respond as we do. Martie used it to explain systematic variation in how and why the sexes differ in their assessments of each other. For example, do men consistently overestimate women's sexual interest in them? As a leading evolutionary psychologist, she has revealed the fascinating story of how sexual attraction is communicated and perceived, often unknowingly. She's studied subtle changes in women's behavior throughout their ovulatory cycle (for instance, ovulating women change the way they dress and are more likely to wear red), and she's always looking to apply the logic of error management theory to new problems.

In an aptly titled review, *The Paranoid Optimist: An Integrative Evolutionary Model of Cognitive Biases,* Martie and her coauthor, Daniel Nettle, start with two contradictory pieces of folk wisdom: "Better safe than sorry," and "Nothing ventured, nothing gained." The first adage represents a conservative strategy, whereas the second is less conservative. Given some estimate of risk, should we be conservative and overestimate risk, or blasé and underestimate risk? Recall Nervous Nelly and Cool Hand Lucy from Chapter 8. These marmots had different decision thresholds. Optimal animals, whether human or nonhuman, should behave in ways that minimize the costs of making mistakes.

Theoretical models suggest that when faced with a starvation-predation risk trade-off and imperfect information about the true risk of predation, being conservative—that is, overestimating risk—may be an optimal strategy. By doing so, animals will minimize the costs of missing a predator. But thinking that everything is a predator, perhaps like Nervous Nelly's alarm calling at a falling leaf, isn't such a great strategy either. It's costly to overestimate risk. In the starvation-predation trade-off condition, you starve to death because you never leave the safety of your burrow or the safety of cover. But how should we determine what is an optimal degree of risk?

As discussed in Chapter 8, consider that many autoimmune diseases are defined by an overactive immune system responding to every challenge as an existential threat. Similarly, but on a larger scale, when a nation over-responds to a threat from another nation, costly wars and quagmires result. There's an optimal amount of time and energy that should be allocated to defense, whether from a furry predator or a predatory nation, and it's necessary to get it right most of the time and in a variety of circumstances. Indeed, one of the mantras of natural security is that we are surrounded by descendants of individuals who got these assessments right.

To test the idea of making assessments in different circumstances, imagine you're walking along a trail in a Costa Rican rain forest. Trees are constantly shedding branches, and as you're walking, you look down and see a long, cylindrical, but not perfectly straight object. Is this a tree branch or a snake? On a street in New York City you'd immediately think it was a stick and kick it out of the way. You'd probably be correct. But in Costa Rica, home to a number of deadly vipers, the consequences of misidentifying the snake are greater. I've experienced this myself, although in a slightly different environment. I was trekking alone in search of gibbons through Taman Negara National Park, the largest park in peninsular Malaysia, when I came across a log in the path. I stepped over it. But as my feet were on either side of it, the log slowly moved forward. It was not a log, but a *huge* reticulated python! I jumped and screamed. Heart racing, I leaped forward. When I turned around, I saw the remarkable serpent slither out of sight into the underbrush.

Thinking sticks are snakes should lead to increased survival. It's the conservative choice and likely successful when the costs of error are asymmetric. An asymmetric cost is one that is quite small—you may scream and run away, but there's no harm in that. If you make this choice, you're unlikely to mistake a poisonous snake for a stick and get bitten. By being conservative, we minimize the costs of making a mistake.

We make many such conservatively biased assessments in our everyday lives. For instance, as we hear a sound get louder we infer that it is approaching us, and thus we systematically estimate the time to impact as shorter than it really is. Similarly, looming sounds are perceived as moving faster than receding sounds. These assessment biases make intuitive sense because we could be hurt if struck by an object without having paid heed to the warning, but the cost of being wrong is slight. Some of our conservative reactions may seem less intuitive. We often back away from those with a grievous wounds or scars, overestimating the risk of infection from people who are injured or whose lesions are not contagious.

Randy Nesse, one of the founders of the field of evolutionary medicine, summarized the idea of cost minimization well with what he has called the smoke detector principle. A smoke detector should be designed to go off when you're (badly) burning your toast as well as when you have a house fire. If the goal is to detect real fires, having an alarm that sometimes responds to burning toast is good because the consequence of missing a house fire is huge (potential death and destruction). The consequence of an alarm that responds to burned toast is relatively small (vast annoyance). By systematically designing smoke detectors to have a more sensitive threshold, manufacturers can assure customers that detectors won't miss a real blaze.

When it comes to making our own decisions about risk, however, we're not always so conservative. In certain situations we overestimate the probability of our success. Martie's work has shown that men often mistakenly believe that women are more sexually interested than they are. And when shown images of neutral female faces, men describe them as illustrating sexual interest. These overestimation biases make evolutionary sense; women are the limiting factor in human male reproductive success. A woman's reproductive success is limited by the fixed number of eggs stored in her ovaries, whereas men have a virtually unlimited supply of sperm. The most productive woman in history had sixty-nine children. The most productive man in history may have been Genghis Khan: about one half of one percent of men have similar genes on their Y-chromosome. Meanwhile, Sultan Moulay Ismaïl of Morocco, "The Bloodthirsty," is reputed to have sired over 800 children. Whether or not these claims are true, there is a huge potential difference in productivity between the sexes. Historically, assistance with child-rearing increases survival of offspring and has influenced women's partner choices. Thus, from the woman's perspective, overlooking a man who will be a committed spouse is inefficient, but partnering with one who is not committed could be disastrous.

By contrast, men can in theory afford to be less discriminating. If a man misses a potential mating opportunity, he misses the possibility of siring a new child. And this lost benefit is a potentially large cost that is substantially larger than the cost of the modest embarrassment associated with being turned down. A psychological mechanism that may explain men's reduced discrimination is the belief that women are interested in them. Humans are full of such self-delusions, as the evolutionary biologist Robert Trivers has articulated so well in his book *The Folly of Fools*. This is certainly not to say that men don't have preferences. There's even an evolutionary benefit to relative choosiness. Disease acquired from a short-term liaison could impact their future fecundity and attractiveness to a long-term mate. But the negative consequences of a short-term dalliance are often substantially fewer for men than for women for one outsize reason: men can't get pregnant. These behavioral biases are well explained by error management theory.

The logic of error management theory can also be used to answer a common question: Why do nations go to war when terrible loss of life is certain? This question, also known as the war paradox, is addressed in Dominic Johnson's book *Overconfidence and War*. Why doesn't one country or both countries back down when there is a crisis? Before 1900, initiators of war tended to win, but since 1900, initiators have lost about half of the time.

Dominic argues that we must understand our perceptual biases in order to control them. It turns out that people believe they are better, stronger, smarter, or better able to solve problems than they actually are. By creating positive self-delusions, we become overconfident. He notes that we must guard against this overconfidence to avoid slipping into wars or other suboptimal political outcomes. Dominic realized that in certain cases, the leaders of countries created a series of positive illusions leading up to the war that they broadcast to their people. These positive illusions systematically underestimated the costs of entering the war. For instance, in World War I the first units of British troops sent to Europe were told that they would be back home by Christmas. Similarly, in World War II, Adolf Hitler discounted the Russian defenses before getting his army trapped in Russia in winter.

The United States overestimated the impact of sending more troops to Vietnam and to Afghanistan.

According to Dominic, rational debate and discussion can temper irrational exuberance. The initial discussions between Neville Chamberlain and Hitler slowed the start of World War II. The Kennedy Administration provides an example of how discussion of options and systematic evaluation made for better decision making. During the Cuban Missile Crisis, both John F. Kennedy and Nikita Khrushchev had close advisors that strongly encouraged them to escalate, which could very easily have led to a nuclear exchange. However, Kennedy and Khrushchev talked to each other about the horrors of nuclear war and its potential impact on their children, and these discussions helped them temper the exuberance for war. Dominic goes so far as to muse that if Kennedy had not been assassinated, we might have been able to get out of Vietnam much earlier.

At one of the NCEAS meetings Dominic and I discussed how the lessons from error management theory had been more or less independently derived in several fields. I'd certainly seen this in my research. In a variety of situations having a conservative bias is indeed adaptive, and animals with such a bias had the highest reproductive success. Dominic and I began a collaboration and a literature review to synthesize these models and insights across disciplines.

Perhaps one of the earliest examples of error management theory comes from the work of seventeenth-century philosopher and mathematician Blaise Pascal. He ponders the question of whether God exists and considers the costs of belief. He argued that if you believe in God but God does not exist, then you will suffer modest costs associated with your belief. If you believe in God and God exists, however, then you would have infinite gains—you would be in Heaven for eternity. The real cost emerges if you do not believe in God and indeed God exists. In this situation you would be punished for an eternity in Hell. Specifically, writing in *Pensées,* Pascal said, "Let us weigh the gain and the loss in wagering that God is. Let us estimate these two chances.

If you gain, you gain all; if you lose, you lose nothing. Wager, then, without hesitation that He is." At least on paper, Pascal convincingly minimizes the cost of a really bad eternity in Hell with error management theory!

There are plenty of recent scientific examples of how error management theory is applied; these examples provide a deeper understanding of why we behave the way we do. For instance, some biologists have hypothesized that allergies flaring from an overreaction to harmless stimuli like pollen may be a side effect of successful protection from parasites and pathogens. Behavioral ecologists have written about the life-dinner principle, which explains why it is better to miss a meal than take a risk that would get you killed. Animals that miss a meal by behaving cautiously might go hungry, but they will likely live another day. Animal behaviorists have used extensively the logic of signal detection theory, which explicitly models the costs and benefits of falsely detecting a pattern versus missing it.

Geneticists have used error management theory to understand whether mutation rates are constant across the genome. The relative cost of a mutation varies depending upon the trait and natural selection, and this variability has selected for different mutation rates to reduce the likelihood that a critical trait will be randomly modified. The cost of lower mutation rates in parts of the genome where costs of a mutation are high is that other parts of the genome, where the costs of a mutation are lower, are relatively more susceptible to mutations. In both of these cases, the likelihood of making a costly error is reduced. All of these recent examples share a common theme: individuals who act to minimize the cost of making a big mistake do better than those who do not minimize this cost.

Even human superstitious behaviors can be explained within an error-management framework. When we're superstitious, we attribute cause and effect when a causal link is really not present. For instance, a friend replaced all of the tires on her Subaru and was then immediately involved in a car accident that totaled it. When I heard this story, I wondered if I was going to be involved in a car accident when I replaced the tires on my car, also a Subaru. The error-management interpretation is that as long as the superstitious event sometimes occurs, and especially with major consequences, natural selection will favor misunderstanding causal relationships in small ways for the potential large payoff acquired. After I put new tires on my car, I drove very slowly and carefully!

The logic of error management theory can be generalized even to organisms other than animals and humans. A number of close colleagues have applied error management theory to explain how plants allocate energy to herbivore defense. Plants face the same problems animals face: how should they allocate scarce energy to defend themselves when doing so will take energy away from growth and reproduction? Plants have a variety of ways to reduce damage or reduce the likelihood of an elk or insect munching on their tasty leaves. But these defenses—which include producing toxic chemicals, making their leaves less palatable, or growing spines to discourage a hungry deer—are costly. Thus, we often see what are referred to as inducible defenses—defenses that are deployed specifically in response to herbivory and not before.

One of my favorite examples of an inducible defense takes us back to the East African savannah. There, giraffes browse on a variety of acacia trees. Acacias have evolved a number of nifty defenses. They grow spines, which provide a physical deterrent to herbivores, and they produce tannins, which are a chemical defense that makes the leaves taste bitter. In addition, acacias provide food and housing for ants. When a branch is disturbed by an herbivore or an interested human, the ants swarm out of their homes and bite the predator. Such defensive mutualisms, when two species do things that benefit both of them, are everywhere. But producing spines is costly, and ideally an acacia would do so only when and where it provided a concrete benefit. Thus, an acacia tree has spines only where (and when) animals eat it. Trees protected from browsing have massively reduced spines, which are ineffective against giraffes and other large mammals. Trees browsed by goats but not giraffes have long spines only on the lower part of the tree, where they are browsed. No tall herbivores? The higher spines are all vestigial. Giraffes around? The long spines grow anywhere a giraffe can reach.

I attended a talk years ago by Truman Young, who has researched inducible defenses. He noticed that in Nairobi, Kenya, the acacias had no risk of giraffe herbivory but nevertheless had spines when they were near the roads. Why? It was the tissue damage of passing traffic that induced the defenses, not the actual herbivory. Thus, acacias in Nairobi, it seems, follow the logic of error management theory: with a rich history of exposure to herbivory, it's better to deploy costly spines than to be eaten. Truman also observed that occasionally a very tall acacia would fall at his study site. The highest branches of these trees—above giraffe height—had produced only the impotent vestigial thorns. On the fallen trees, these defenseless branches were now within reach of the voracious giraffes, which quickly stripped them of all their leaves.

Thus, whether we are looking at plant, animal, or human behavior, error management gives us the tools to understand adaptive responses to threats, whether superstitious, philosophical, or all too real. Organisms attempt to minimize the likelihood of large, costly mistakes.

Error management theory and the power of positive illusions have gotten us into a pickle with respect to actions we should be taking now to reduce human-driven global warming, however. The immediate costs of acknowledging the reality of global warming would be significant. Thus many of us would prefer to ignore present pain for much greater but more distant consequences. I discuss why climate change is such a complex issue in Chapter 11, but here we'll look at what's really at stake.

The global scientific consensus is that climate change is driven by human use of fossil fuels. The consequences of this heating are profound and widespread. As the Greenland and Antarctic ice caps melt, more water flows into the oceans. With more melting of polar sea ice in the summer, the ocean water warms because a dark object (the sea) captures more heat than a bright, reflective object (ice). The resulting increase in sea level and changes in global ocean circulation patterns will change weather throughout the world. We're already experiencing increasing storm intensities because there is more moisture in the air. The Atlantic hurricane season of 2017 included seventeen named storms and ten hurricanes, six of them in Categories 3, 4, and 5. Between 2014 and 2018, the world experienced on average the five warmest Septembers on record. Increased heat also will render some parts of the world uninhabitable because there are upper limits on our ability to grow food or live well in extreme dry heat. Low- and lower-middle-income countries are predicted to suffer the most. Leaders of many nations, and even religious leaders such as the Dalai Lama and the pope, have acknowledged the science indicating global warming. Pope Francis, in his 2015 encyclical letter "*Laudato Si*: On Care for Our Common Home," has called upon humanity to come together and work toward a solution.

The debate about how we should respond to this existential threat should focus on action. This becomes a bit problematic because we have no well-developed ethics to guide us through what is really an intergenerational problem. For instance, how much should the current generation's growth be curtailed in order to protect future generations? The Haudenosaunee Confederacy (Native American Iroquois) believe that decisions should be evaluated by considering the consequences seven generations in the future. Putting such value on future generations at the expense of current generations is worthy of a healthy debate. Industries with a lot to lose (coal, petroleum) have cast doubt on the evidence in some disingenuous ways that follow the playbook created by tobacco lobbyists in the 1950s and 1960s.

Can technology save us from the deleterious consequences of climate change? Pointing to huge increases in energy conservation technology that have already reduced carbon emissions, some wonder why we should pay a cost now if future, as yet undiscovered technological innovations may help us better pull carbon out of the atmosphere or create other ways to cool the Earth. I discuss this possibility in more detail in Chapter 11.

In October 2018, the United Nations Intergovernmental Panel on Climate Change issued an alarming report. They argued that humanity had only about thirty years to work together to create what they call "net zero"' carbon production—where any carbon released into the atmosphere is balanced by that extracted from the atmosphere. They argued that massive conservation and life-changing actions are needed immediately. By doing so, warming could be limited to 1.5°C above preindustrial levels. Exceeding this warming would, they warned, wreak havoc on ecosystems, human infrastructure, and humanity. But it will take a lot to change our lives so quickly. The next generation of young people, who will have to live with these changes, has recently been prominent in advocating for change through school strikes in 2019. Perhaps they will encourage other sectors of society to engage. After all, our fitness depends on leaving descendants, and those descendants must suffer the consequences of our actions long after we are dead and buried.

Given what we've learned about error management, I'd argue that the consequences of inaction with regard to climate change are so great, and the chance of truly game-changing, unknown technological innovations are so low, that we should do everything we can to reduce anticipated costs now. This is also known as the precautionary principle. Better safe than sorry.

11

OUR INNER MARMOT

Throughout this journey we've seen how in the wild fear motivates prey to respond to a variety of cues (as discussed in Chapters 2, 3, and 4), to be aware and employ sophisticated economic logic to minimize the chance of becoming someone's meal (as discussed in Chapters 5 and 6), to learn quickly at times (Chapter 7), and to seek information about threats not only from others of your own species but from other species (Chapter 8). Animals, like us, are biased in their risk assessments, and error management theory (Chapter 10) explains why.

In this chapter I'll apply some of these lessons from animals to discuss the human relationship with fear. I acknowledge that for some this may be a leap too far. Critics might ask if we can really argue that a marmot's response to a predator is similar to our own anxieties and fears about complex social, environmental, and political events. I suggest turning this question around. Why invoke human exceptionalism when we share the same neurophysiological hardware with many other species? We can identify similar responses in animals and in humans to a variety of stimuli. However, I acknowledge that I'll be going a bit further in these chapters as I develop testable hypotheses about how we can harness our fundamental understanding of animals' response to fear-inducing events to help us live better with fear. I'll ask if we can apply lessons from life to improve our decisions about things that induce fear or provide existential threats to our well-being.

In this experiment I want to know how much we can learn from our inner marmot. To explore this we will use our growing intellectual toolkit about fear to address some major human problems, and in Chapter 12 we will distill the lessons from animal and human risk assessment into a series of lessons by which we can better thrive in a risky and potentially fearful world.

For humans, we find that fear motivates change the best when the message is simple. It helps if the potential outcome is disgusting; disgust is a powerful motivator. In one of my favorite examples, officials were struggling to combat the rise in methamphetamine abuse. Because meth is so highly addictive, public health informational campaigns had limited success. Enter Tom Siebel, a software millionaire on a mission. For his campaign, Siebel focused on the dental decay and meth-mouth condition associated with meth abuse, an excellent example of how fear and disgust motivate people. Starting in 2005, he supported a highly successful advertising campaign of highway billboards in Montana. These billboards showed close-ups of rotten teeth in a woman's face labeled, "YOU'LL NEVER WORRY ABOUT LIPSTICK ON YOUR TEETH AGAIN." Then he created short commercials that showed the horrible futures of unsuspecting users, including young women selling their bodies to make money to support their habit. The campaign worked—high school meth use dropped 45 percent in two years. We can compare this result to the 7.8 percent annual drops before Siebel's shock campaign started.

By contrast, complex problems where the causality is less direct require different motivators. In part this is because fear most successfully leads to immediate responses, and there are substantial costs to maintaining a fearful state for too long. As all successful animals know, you can't hide from threats forever. You eventually have to come out to forage. Further, we can't constantly deploy our fight-or-flight mechanism without deleterious health consequences, like stress-induced illness. Also, using fear to motivate change has an important shortcoming: we are likely to habituate to fearful messages.

During the marmot active season, much of my time is spent sitting in meadows observing marmots. Marmots are not as socially interactive as monkeys or meerkats, so social interactions are comparatively rare. For this reason we scan the hillsides and meadows to keep track of the marmots and predators in and around our colonies. When I was at my prime, I prided myself in frequently detecting predators before the marmots did.

We also keep track of marmots emitting alarm calls and the marmots' response to callers. Most marmots emit a single alarm call. We look up to see a meadow full of poker-faced marmots, rearing up erect on their hind legs, looking intently at something. But what are they looking at? We often fail to identify the source of their concern. But when marmots emit more than a single call, we count the alarms and can often identify the caller and the stimulus. The stimuli vary, of course. Coyotes and foxes can elicit multiple alarm calls, but sometimes marmots call for over an hour in response to the sight of a single deer. Once, an exhausted assistant came back to the lab stating that she had just counted 1,876 alarm calls!

In Chapters 7 and 8 we discussed habituation and the value of reliability assessment. Recall Cool Hand Lucy and Nervous Nelly. We know that hungry marmots can't simply stop everything for an hour because Nervous Nelly is calling in response to a deer. Wise individuals should habituate to nonthreatening, noninformative alarm calls and eventually resume their daily business. The same thing that drives animals to habituate, to reduce the costs of constant vigilance, occurs in humans as well.

In a working group of the National Center for Ecological Analysis and Synthesis (NCEAS), we speculated that habituation reduces responsiveness to fearful messages. Specifically, we wanted to know if the Department of Homeland Security's threat level could ever work with the public given the theoretical expectations from the theory of habituation. Initially designed for first responders, the threat level ranges from 1 (low) to 5 (severe) as a rough assessment of the probability of a terrorist attack. Yet, for about the first five years of its use, the threat level remained either at level 3 (elevated) or 4 (high). For levels 3 and higher US citizens were encouraged to visit the DHS website (www. Ready.gov) for information to protect themselves. (The site has changed its content substantially since the early post–9-11 years.) But did they? If Americans habituated to threat levels, the policy outcome could be opposite what was initially desired. If the threat remains high for a long time, people will let down their guard.

Using polling data, we found some evidence of habituation to the 9-11 terrorist attacks; people's perceptions about risk declined over time. This makes sense from a Bayesian perspective because there were no immediate follow-up attacks, and constant vigilance is costly. Interestingly, when we looked for a relationship between perceptions of risk and the DHS threat level, we found no strong evidence that American citizens responded to increases in threat level by increasing their perceptions of risk. Additionally, the number of page views to www. Ready.gov and phone calls to 1-800-BE-READY decreased over time and were not sensitive to variation in the highest DHS threat level that month. Statistically, after accounting for variation explained by month, threat level had no significant effect on page views or phone calls. This response suggests that people had habituated to the warnings.

When fear is used as an agent of change, we should expect habituation. Indeed, designing habituation-proof stimuli to scare away "problem animals" and reduce human-wildlife conflict is a compelling challenge. The plastic owl above the outdoor urban park dining area worked for a few days before the pigeons decided it created a nice windbreak. And the airport air cannons used to scare birds away from runways work for a while. But animals will habituate to the same present or repeated stimulus as long as there is no negative consequence associated with the stimulus. Thus, mixing up stimuli and modalities to reduce redundancy and tap into different perceptual systems is the best practice to delay habituation.

How can we translate these insights from nature to help us better solve human problems? One promising way to delay habituation may be to have one main message presented in multiple ways. Montana's meth-mouth campaign had one simple message but generated a billboard and a number of different commercials to deliver the message. Tapping into multiple modalities—capitalizing on the sound and sight of fear—made more compelling advertisements and potentially delayed the inevitable habituation.

Simple messages that tap into our fears can be so effective that they may result in wars! Consider the claims that Iraq under leader Saddam Hussein had weapons of mass destruction. If we could only eliminate them, the messaging indicated, we would be safe. We see how poorly that turned out. I wrote the original version of this section in the heat of much posturing by President Donald Trump and Supreme Leader Kim Jong Un over North Korea's nuclear program. And while later meetings appeared to open communication channels about denuclearization, in 2019 North Korea resumed ballistic missile testing. North Korea's simple message, that it will continue testing, successfully taps into our fears. Detonated nuclear weapons, apart from causing widespread immediate destruction and long-term human and environmental devastation from radiation, fill the atmosphere with soot. The effect is not simply local and would likely plunge the entire world into a nuclear winter that would cause global crop failures and famines. Politicians flippantly threatening to use nuclear weapons is almost too much for me to process—a truly effective fear stimulus.

By contrast, complex problems, where the causality is less direct, require different motivations. In part this is because fear most successfully leads to immediate responses, and there are substantial costs to maintaining a fearful state for too long. As all successful animals know, you can't hide from threats forever; you eventually have to come out to forage. Similarly we can't constantly deploy our fight-or-flight mechanism without having stress-induced illness. Yet, in humans, fear is about much more than simply getting killed. I wonder if fear or anxiety could work in other social domains to motivate change?

But what happens when technological advances reduce risk? Do we revel in our increased security? Strangely, it seems that we don't. Rather we engage in what's called risk compensation, we paradoxically accept more risks in response to technology that reduces risks.

It is certainly much safer to drive now than 1965 when Ralph Nader declared the Chevrolet Corvair "unsafe at any speed." Cars now have better suspension and brakes to keep you on the road. Manufacturers have designed cars with multiple safeguards to prevent you from shooting head first through the windshield should you stop abruptly, including seat belts and air bags. And cars now have antilock brakes and collision avoidance systems. While the jury is still out on collision avoidance, lessons from when airbags and antilock brakes were first released are revealing.

When passive restraint systems were first introduced, some people stopped putting on their seat belts because they assumed that they would be safe with airbags alone. In reality, these systems work together to increase your odds of surviving a bad accident and keep you in the car, where you typically are safer. There are studies that have shown that taxi drivers drove faster on wet surfaces when antilock brakes were first introduced because they could drive faster with a reduced risk of hydroplaning. However, driving faster on wet streets, even with antilock brake technology, isn't really a great idea; there was no net reduction in car accidents.

The safety of much of the technology we use to get to work everyday but also to have fun on vacation has improved. Skiing, for example, has never been safer. I remember skiing with bindings that would release from side to side if you leaned too far forward but not if you leaned back. The outcome was inevitably poor if you fell in a certain way. Now, ski bindings are much more predictable and efficient—releasing when you need them to in all directions but not when you don't want them to—when speeding down an icy slope.

In the last ten years we have seen a revolution in ski design. I joke that new ski design gives you about ten more years of skiing because it makes it easier to turn quickly with tired, old legs. Easy turning makes it easy to stop quickly. While most skiers did not wear helmets when I learned to ski over fifty years ago, almost all do today. Skiing with a helmet is absolutely safer. It often (but sadly not always) provides protection against traumatic brain injuries. By some measures, skiing is getting safer, but by others, helmet use is not reducing brain injuries. This may be because skiers are using this better, safer equipment to push their limits. The unfortunate ones go off cliffs or hit trees—catastrophic injuries that safer equipment can't fully prevent. Ski fatality rates are rather flat, at 1.06 per million skier days.

The same ski improvements that make it safer to ski on groomed trails have led to an absolute revolution in backcountry skiing. Safer, lighter equipment and better avalanche gear have led to more people in the backcountry seeking powder on ungroomed slopes. And with more people in the backcountry, more people are getting caught and killed in avalanches.

Foolproof: Why Safety Can be Dangerous and How Danger Makes Us Safe is a wonderful book by Greg Ip. He writes about how trying to make things safe encourages people to accept greater risk. His examples range from federal legislation following an economic crash to catastrophic storm insurance. For instance, when a hurricane flattens a community or a flood swamps it, the immediate response is to rebuild rather than to reconsider whether it's safe to live there. Building a levee is a great way to reduce the risks associated with living in a flood-prone area such as the Netherlands or the city of New Orleans, Louisiana— until it breaks. If there was no levy, then nobody would be exposed to risk because nobody could live there. Additionally, government-backed insurance encourages people to live in locations that no private insurance company would insure, or would ensure only at a prohibitive price. But insurance reduces our anxieties about losing all of the capital we've invested in our homes. We often don't stop to think about why some insurance is so expensive. The comfort level even costly insurance provides emboldens us to rebuild in risky areas, even if it is likely that we will face the same problem in the future, whether the risk is rising sea levels or wildfires. Thus, our perceptions of safety counterintuitively encourage risky behavior. The opposite is true as well: if we perceive something to be dangerous, such as commercial airline flight, we will do whatever it takes to be safe. In consequence, consumer demand, bolstered by federal regulation, has made commercial air travel remarkably safe.

All of these examples illustrate how we often fail to reap the benefits of reduced risk and improved safety. When we think we are in control, we are in fact overconfident. But I think risk compensation is not exclusive to humans because I see the same process occurring in marmots. Marmots that run relatively slowly are also warier when they forage. By doing so, they behaviorally compensate for their physical limitations. Marmots that are socially isolated, as I discussed in Chapter 8, are more likely to emit alarm calls that could function to dissuade predators.

Fear, while not a uniquely human attribute, *makes us human,* and managing it is an inevitable part of life. Not only is it impossible to totally eliminate risk or the fear and anxiety that accompany it, but we seem to have a *desire* to maintain risks and challenge ourselves even when we can live safer lives.

Politicians know that capitalizing on fear wins elections or earns support for laws. As discussed in Chapter 3, President Lyndon B. Johnson's *Daisy* advertisement, which portrayed a child counting daisy petals along with a countdown to a hydrogen bomb explosion, was compelling because it incited fear. The lasting images in that advertisement promised safety through a vote for Johnson.

And, let's not forget how President Bill Clinton passed his 1994 federal crime bill with First Lady Hillary Clinton's help. Hillary referred to young criminal gang members perceived as remorseless as "super predators." This was in the aftermath of the 1992 riots in Los Angeles and at a time when the United States was in the midst of a crack cocaine epidemic. Urban gangs were seen as violent and scary, and she was clearly referring mostly to black and Latino youth. By using the term in speeches given to white audiences, she effectively capitalized on their implicit biases and fears to drive support of the bill.

Fear sells, and this message is not lost on either Republican or Democratic strategists. Fear can be used by politicians to gain reelection or pass bills, but it can also be used to logically manage our risks and work to reduce them. There are immense social and public health issues raised by living around risks. Wise government, whether local, state, or federal, is able to develop policies to identify and manage risk and respond quickly and adaptively when the worst happens, whether or not the event can be predicted. Fear should motivate us to design resilient systems.

At our Natural Security working group meetings we realized that good defensive systems are adaptable. It would not be efficient to create a detector for each potential bioweapon—anthrax, smallpox, plague, and so on. Where would you put them? How much would they cost? Would they be in places where a pathogen was likely to be detected? Like the Maginot Line, built by the French after World War I, any fixed defense could be easily avoided. It quickly became clear to us that creating specialized detectors was neither efficient nor desired. Instead we argued for creating adaptable, flexible, and multipurpose defenses. For instance, a good public health system can efficiently detect and communicate to everyone the outbreak of symptoms associated with a bioweapon attack in time for effective public health measures to be deployed.

Similarly, civil defense requires flexible governmental agencies, and we've seen that the responses of government bureaucracies can be sclerotic. The Cajun Navy, a community group that emerged in the aftermath of Hurricane Harvey in 2017 offers a promising model. Immediately after the floods, people from adjacent states began showing up with their boats in tow and to help rescue victims. The Cajun Navy was born, and is now a registered 501(c)3 organization. Underequipped government agencies tasked with rescuing people generally welcomed this adaptable response.

Evolution, as the Nobel Prize–winning biologist François Jacob noted, is about tinkering with what you have rather than making up new solutions to new problems. Thus, about 250 million years ago, when the first termites evolved sociality, this same suite of neuroendocrinological responses formed the basis of responses to social threats—the threat of losing a meal, a territory, or the all-important status that comes with social rank. As discussed in Chapter 1, the suite of neuro-endocrinological responses—the same ones that give us sweaty palms, a racing heart, and wide-open eyes when we narrowly avoid a car accident or a potentially catastrophic fall—ensured that animals avoided risks when detected and, when challenged, prepared themselves for fighting and fleeing by shunting energy and focus to predator evasion. This is our evolutionary legacy; it predicts how we respond to threats and predisposes us to overreact to a variety of modern threats our animal ancestors never dealt with. And these threats, or at least our perceptions of them, are abundant. We have a long history of intergroup conflict that predates the evolution of modern humans. We fear outsiders, meaning anyone not from our own group. We fear violent interactions with potential predators, human or otherwise. We fear the loss of our resources and the implications of this for our own security and our family's security.

We have not had sufficient time as a species to tune our responses to a variety of novel threats. Rather than habituating to this constant exposure to both historically important and novel threats, we have become fear conditioned. These limited experiences with bad events form indelible memories that prepare us for defensive action. Fear conditioning may be the basis of post-traumatic stress disorder, which afflicts civilian victims of violence and military personnel. Indeed, like marmots and many other animals, humans seem to be primed to learn to avoid violence and then remember these lessons for a long time. In general, when a trait evolves under one set of circumstances and those circumstances later change, the trait may no longer be adaptive, an evolutionary mismatch.

Our ability to learn about potentially bad things from a single experience would be predicted from error management theory: better safe than sorry. Thus, fear conditioning is expected and likely very adaptive. With a twenty-four-hour news cycle pumping out fearful messages, however, fear conditioning may be debilitating. Our exquisitely evolved fear system results in a desire to eliminate risks, however small they really are. But of course it's impossible to eliminate all risks; it's impossible to eliminate all fears. It's not even desirable. We are successful humans because our ancestors' fears allowed them to survive.

An astute reader may notice that I discuss habituation in conjunction with the idea of using fear to motivate behavioral change but sensitization when discussing the twenty-four-hour news cycle. Both are possible outcomes. I believe that it's remarkable, given over a century of formal study of habituation and sensitization, that we really do not have a good idea of the natural history underlying these processes. In other words, we don't know which response to expect in reaction to many natural circumstances. Continued study of diverse species in a variety of natural situations will generate the data necessary for better predictions about the conditions under which habituation and sensitization occur. This knowledge will help us plan strategically for future human-made or natural disasters.

On a personal level, you don't need to wait for better data about habituation and sensitization. Although risk is unavoidable and ever-present, you can manage your own risk by pausing and evaluating the data rather than reflexively enduring a fear-conditioned response. Let's look at a few issues that cause or have caused anxiety in the public lately. Many Americans fear outsiders, in particular those who have immigrated to the United States recently without papers. The vast majority of "illegal" immigrants are not murderers or rapists, as some American politicians and media outlets have suggested. Like many of our immigrant ancestors, they are simply seeking out a better life for themselves and their families. The vast majority of refugees fleeing wars are not intent on forming sleeper cells and attacking the homeland but are seeking out a safer and better life, much like those who escaped Nazi-occupied countries before and during World War II. And the vast majority of those with strong religious beliefs are not religious terrorists but people who find solace and guidance by congregating with those with similar beliefs without persecution, much like some of the earliest immigrants to the United States.

Understanding the biological basis of our fears may prevent our intellects from being hijacked by political agendas. Letting others dictate the nature of our fears respects neither our fellow humans nor the exquisite evolutionary legacy bequeathed to us.

So what advice might our inner marmot give for our current existential threat—climate change? Our inner marmot has evolved to address simple causal links, such as, "If I see a predator in a particular location on several different occasions, it probably means that this is a risky place." While our inner marmot is hungry, she will probably forego eating in that risky location, and so live and eat another day. We have also seen how this works in the bushbuck in Gorongosa National Park. Once predators were essentially eliminated from the park, the bushbuck left the safety of the dense forest and began to forage on higher quality food in more open areas. The trouble is, climate change, unlike the immediate problem of figuring out whether it is safe to eat, is complex in ways that befuddle our desire for simple causality and hence simple solutions, as Steve Gardiner notes in his book *A Perfect Moral Storm*. Why?

First, our inner marmot can't see the effect of any individual action. Our individual actions on climate change today are hardly visible given the magnitude of the problem. It really isn't how much carbon I personally use or how much carbon you personally use. The scale of the global carbon cycle is immense and likely beyond our evolved cognitive understanding. We convince ourselves that our individual actions have a limited effect, so we should be excused from changing our individual behavior—a phenomenon connected with the well-known "tragedy of the commons," in which a shared resource is spoiled by the self-interested actions of its members.

This is not to say that individual actions can't have an effect; they can if widely adopted. Generating such adoption, if not forced, will require a common understanding of the collective impacts we create as well as a solution that is preferred. Taking away entitlements or perceived entitlements—such as most Americans' belief that we all are entitled to a private automobile—is much more difficult than incentivizing change. A large psychological literature has shown that we are averse to real and perceived losses and costs, and we undervalue gains and benefits. To make an impact on a large-scale social problem, people have to *want* to change their behavior. They have to believe that change is better for them. Thus, the challenge.

Second, our collective actions today—such as burning fossil fuels—will not have immediate consequences. The sum total of what we've done in the past, what we do now, and what we will do in the future, however, will have huge consequences for future generations. As Steve Gardiner notes, we simply don't have a set of established ethical principles for dealing with such intergenerational problems. It's clearly unacceptable to cause pain and suffering today to reduce pain and suffering for those who are unborn. But it's also equally unacceptable to destroy the hopes and options of others in the future because of our actions today. Where you fall on this continuum will influence what you find acceptable—and how much you will discount the consequences of our actions today. In the spring of 2020, American governors and mayors pleaded for just this sort of collective action. With yet-silent COVID-19 infections on the rise and demographic evidence from China and Italy available, states and cities urged their residents to stay home to slow the spread of the coronavirus and to prevent hospitals from being overwhelmed in the coming weeks in their states and cities. Even though the consequences of behavioral change were not immediately visible, most people could accept the message that staying in their homes could help save lives of those at risk. The climate change issue, however, presents a stern challenge for our empathy and imagination.

Third, we could stop all carbon use tomorrow, and we'd still have ongoing climatic warming. This is because atmospheric carbon dioxide doesn't simply break down immediately. It persists in the atmosphere until it breaks down and continues to keep heat from radiating out to space, acting like the glass in a greenhouse. Methane, released when natural gas burns, is an even more effective greenhouse gas, but it breaks down faster. The problem will persist until the greenhouse gasses break down or until they are extracted from the atmosphere. The persistence of these gasses means that even after we attempt to fix the problem, it may get worse before it gets better, further complicating our ability to draw causal relationships from our actions.

Fourth, because the warming that we've started is nonlinear, it's harder for us to imagine. In nonlinear systems a unit change in one variable does not necessarily cause a unit change in a response. Turn up the volume on your stereo and the music gets louder—to a point. Once that point is reached, further increases in volume lead to all sorts of noisy distortion. The system has entered its nonlinear phase. Consider the melting polar seas. As long as there is summer sea ice, the reflectance off the ice prevents the water from absorbing solar energy and warming. Add more heat, lose more ice. Once sufficient ice melts, the ocean will be warmed by the extra solar energy absorbed by the dark sea surface. In other words, the rate of ice melt will increase. The result? Summers will be ice free, the sea will get even hotter, and there will be changes in the global currents in both the air and the sea that determine so much of our weather. Indeed, oil companies have begun exploring oil extraction in seas that until very recently were covered by ice almost the entire year. Cargo companies are exploring the ability to ship cargo though the highly dangerous but lucrative Northwest Passage. President Donald Trump, failing in a 2019 bid to purchase Greenland, has proposed a consular outpost on the strategically located and mineral-rich island, perhaps in part to guarantee access to what will likely be ice-free shipping and newly exposed natural resources. Tourists in the remote north and south have begun to go on summer voyages through areas traditionally filled with ship-crushing ice.

There is simply no single solution to the problem. Plausible visions for the future include increased energy conservation through increased efficiency, reductions in energy use, reductions in the consumption of animal products (especially red meat, milk, and cheese), increased use of renewable resources like wind and solar, increased use of nuclear energy (which, once operational, emits no carbon), and removal of carbon dioxide from the atmosphere through a variety of techniques (none of which have yet been plausibly scaled up). Easy problems are those with simple and apparent solutions. Wicked problems are those that resist resolution because of their complexity. Slowing our anthropogenically driven march toward higher temperatures, more extreme storms, and massive biodiversity loss is truly a wicked problem. Our inner marmot is unprepared.

So, what's the message? Will fear be part of the solution to motivate our behavior to address climate change? There's considerable debate on whether doom-and-gloom environmental messages work.

In general, I expect that simple problems that have simple causal pathways leading to simple solutions might be solved by messaging that provokes fear. For instance, to evacuate people ahead of a hurricane, civil defense authorities and political leaders resort to fear—comparing the current storm to past storms and highlighting how much damage was caused in the past. They've even been known to make it personal by instructing those who do not want to leave to write their social security numbers on their arms with markers and to provide first responders with names of their next of kin before the storm. Seismologists have started to shift their messages to those near fault lines, pointing out that it's not if a catastrophic earthquake will strike but when. Such messaging capitalizes on our fears in order to increase the likelihood that people are prepared. Fear of hospitals being overrun by an influx of critically ill patients, like what occurred in Italy in March 2020, led to some regional successes in rapidly changing people's behavior in the early days of the world's response to the COVID-19 pandemic. Over-strained health professionals who shared stories and images worldwide personalized the health crisis, leading to action.

Even if the scale of the problem is huge, as long as the causality is straightforward, fearful messages may work. But we must be wary of artificial simplicity and messages that intend to promote fear in more complex situations; the messages in these scenarios may backfire.

12

WISELY LIVING WITH FEAR

As we've seen over the course of this journey, the most essential and fundamental challenge of life is to strike a balance between taking risks and staying safe. We've learned why animals are appropriately cautious and why we should be too. But this ancient existential balance requires equal consideration of another stark truth: those who are too cautious will either starve or be outcompeted by those less cautious. Those who are too cautious will not survive.

We need to know both what to fear and what not to fear. We have a suite of specialized cognitive abilities at our disposal that themselves were the product of natural selection. We must integrate this specialized knowledge with the intellectual toolkit and framework we've developed by studying our successful ancestors. Formal cost-benefit analyses are often used by humans to make decisions about whether or not a risk is worth it, and these analyses are routinely used to inform policy decisions. But our evaluations of risk are often biased: they are swayed by context, history, likelihood, and expected payoffs. Understanding how we make decisions is essential for us to live wisely with fear.

We are a remarkably illogical species, and our decisions are not solely based on quantitative estimates. Why do we fear nuclear power when more people are killed by the process of transforming coal into energy, and many more are at risk from the climate disruption that results from our profligate use of fossil fuels? Why don't we worry as much about driving fast when we wear seatbelts or have airbags installed in our cars? We overestimate our ability to get ourselves out of problems and under-plan how to avoid getting into them in the first place. So what should we fear? And how should we make informed risk assessments about these fearful things?

A simple way of viewing risk is to consider two factors: 1) the probability of something happening; and 2) the consequence if it happens. A very rare event with a modest consequence may not be that risky, while a very rare event with a huge consequence is something to thoughtfully consider. For example, contrast the risk of slipping in the shower with the risk of a nuclear war. As we age, the consequences of slipping in the shower increase, and in older people, a broken hip may lead to hospitalization, infirmity, and even, on occasion, death. Thus, it's reasonable that our perceptions of the risk of taking a shower change as we age. We are more careful entering and leaving the shower. We may install a nonslip mat. By contrast, the risk of a nuclear war may be small, but the consequences are profound and horrific, likely affecting everyone on Earth. But except in the most straightforward cases, the exact probabilities of many events are unknown, and the consequences are a function of many variables.

To properly calculate risk, we must collect data. Then we must assess the certainty of our estimates. For instance, if many deer are hit by cars and trucks in a particular section of a busy four-lane highway, there are a number of things that can be done to reduce the risk of injury and death to deer and to humans. Signage can inform people to be aware of deer crossing the road. Speed limits can be reduced. Deer-proof fences can be installed to keep deer off the highway. If we're really motivated, we can create overpasses or underpasses for the deer and other wildlife. Such wildlife crossings have been very effective strategies for reducing mortality in migratory populations of pronghorn antelope and other species. With time and with data, we become more certain of where risk mitigation should be deployed. Our certainty on a rural road will be diminished because we have less plentiful information; fewer people travel on these roads.

To help calibrate risks and better understand the underlying factors that drive our decision making, I offer fifteen principles of risk assessment. Some are generated from insights we've learned throughout the book, and others have emerged from the more specific study of human decision making.

1. *Our perceptions of our own mortality risk change as we age.* For example, as discussed above, the consequences of slipping in the shower increase as we grow older, and we may compensate for this increased vulnerability. By contrast, we know that teenagers do a variety of risky things that increase their risk of injury and death. This explains why your insurance goes up substantially when you add a teenager to your list of drivers. Common wisdom infers this result from teenagers' sense of immortality. Yet, data suggest the opposite. Rather than thinking that they will live forever, teenagers overestimate the likelihood of dying in the next year. Given this perspective, it may seem acceptable for them to take larger risks if there are commensurate benefits to those who take risks—nothing ventured, nothing gained. Perceptions of longevity influence our decisions about accepting risks, and these vary by individual and by age. Growing up in a dangerous neighborhood with high infant and childhood mortality may encourage a live-fast-and-die-young mindset. By recognizing these changes in vulnerabilities and perceptions, we can make better decisions that are based on the true costs and benefits of a potentially risky action or activity.

2. *Our decisions may be influenced by our pre-existing beliefs.* We, like other species, learn in a Bayesian way—past experience matters! And once we get something in our mind, it's difficult to shift it out. For instance, if we are debating whether or not guns make people safer, data can be interpreted in ways that support or reject the need for and morality of personal gun ownership. Consider that between 2010 and 2015, the United States was ranked fifty-ninth in global homicide; there were 2.70 murders per 100,000 people annually. Honduras was ranked first, with 67.19 homicides per 100,000 people per year. However, Honduras only had an average of 5,218 gun-related murders annually

while the United States had 8,592. Brazil, ranked twelfth, had the most murders annually—38,494. Given these numbers, there's both room for interpretation and room for rationalization based on your pre-existing beliefs about whether citizens should or should not possess guns.

3. *We, like other species, hate to lose.* We are more concerned about losing something small than excited about potentially gaining something large; we made more upset by a pay cut of $100 per week than happier by a raise of $100 per week. Loss-aversion bias, as this is called, is associated with the amygdala—the part of the brain responsible for managing our fears. Remember that the amygdala is directly activated by fearful stimuli, including a predator's scent. By studying two women with an exceptionally rare genetic disease that damaged their amygdalae and comparing them to healthy subjects, researchers found that all had the same ability to estimate rewards. However, those patients with damaged amygdalae showed no evidence of the loss aversion that was present in the healthy subjects. These results match those found in studies of nonhuman primates, in which the amygdala has been suggested to be associated with loss aversion. We *fear* loss.

Marketers capitalize on this loss-aversion bias. Marketers manipulate their messages to focus on how we can avoid potential losses by buying a certain item without necessarily discussing the potential risk versus benefit. Insurance sales capitalize on this fear. Political lobbyists and those trying to influence policy also carefully word their messages because they know that how a risk is presented may change our perception of whether something with some risk is considered acceptable or not. To inoculate yourselves against this, reframe statements about loss to reflect the potential gains associated with a policy or a purchase. You may very well decide to try to avoid losses anyway, but at least you will have had the opportunity to consider the magnitude of the associated benefit.

4. *Our assessments of risk are influenced by whether we accept the risk willingly or whether a risk is being imposed on us.* We accept the risk of injury or death while skiing but often shun risks that our employers may make us accept. When we accept a risk, we also are sensitive to

the benefits we are willing to accept. For instance, it's fun to ski, and fun is its own reward. Similarly, we see that animals accept the risks associated with play (which takes time and energy and may increase the risk of an injury) because the benefits (improved motor skills, neurogenesis, and better condition, among others) presumably outweigh the costs. But play, in a proximate sense, is fun, and fighting is risky and scary, even though play may employ the same actions and movements as fighting. Because getting hurt at work is not high on most people's list of fun things to do, risky jobs often require greater financial compensation to offset the risks.

Commercial fishing and crabbing off Alaska is highly profitable—if you survive. The oceans are rough, and ships occasionally disappear without a Mayday call. Many people are swept overboard or maimed by heavy equipment in a dynamic, slippery, and cold environment. Given the relatively small size of the commercial fishing fleet, it's one of the most dangerous jobs on the planet. The US Centers for Disease Control and Prevention concur. Between 1992 and 2008 there were 128 deaths per 100,000 commercial fishermen, compared with four per 100,000 workers for all other jobs. Commercial fishing off coastal Alaska was slightly more dangerous than commercial fishing in the northeastern United States—the second most hazardous place to fish in the United States. On Alaskan fishing boats, the hazards must be compensated for. A captain on a commercial crabbing boat can take home $200,000 a year, and crew may earn as much as $100,000 per year for only a few months of very hard and dangerous labor. This increased compensation further illustrates that we're sensitive to both costs and benefits as well as the amount of control we have over risky situations.

5. *The type of outcome influences what we fear.* We are more likely to fear dreadful, horrific outcomes. It's likely difficult to induce too much concern about the risk of having yellowed toenails, which might at most influence your status at the fitness club, unless you're a supermodel, and in that case you probably have an insurance policy. The risk of a traumatic, bloody amputation, however, will capture most

people's attention. All of us wish to avoid major, physically traumatic events.

Boeing should have known this when their new 737 Max airplanes began falling from the sky. Airline crashes have a special place in the news: we're captivated by them. The real causes often take time to sort out, but each crash immediately attracts rampant speculation and discussion. Why? A commercial airplane crash taps into our pre-existing biases. Crammed like sardines in a pressurized tin hurtling along at nearly the speed of sound, we have no control when we fly. The outcome of hitting the ground (or the sea) at over 500 miles per hour is simply horrific; planes and bodies are pulverized into small bits. Despite these extreme and extremely rare outcomes, modern commercial air travel is exceptionally safe, and the widely cited statistic that it's safer to fly on a commercial jet than to drive to the airport is true. Nevertheless, maintaining trust in the regulatory system that certifies planes, plane parts, and pilots is essential, and any rumor hinting of corruption or poor quality control will immediately be seized upon.

It is possible that Boeing could have mitigated some of the fear surrounding two crashes involving the Boeing 737 Max in late 2018 and early 2019 by being completely transparent about their investigation with the public and with the US Federal Aviation Administration and other regulatory agencies around the world. Had they voluntarily called for the immediate grounding of the fleet until the problem was solved, they might have avoided losing trust. The last thing any company wants is to have their product generate fear and anxiety in its customers.

6. We are more likely to fear the unknown than the known. This likely explains why nuclear power plants are feared even though there have not been any deaths associated with nuclear energy in the United States since the Shippingport Atomic Power Station began operation in 1957. By contrast, between 1999 and 2015 alone, there were over half a million gun deaths in the United States. Guns are the number twelve cause of death overall and the number one cause of death by homicide and suicide in the United States. An average of 33,400 people die by guns annually. Yet, at least as viewed through the lens of policy, we fear nuclear power plants more than guns.

As we learn more, we fear less; education can inoculate us against false fears. When HIV was first recognized to be a fatal and transmissible disease, HIV-positive patients were stigmatized and isolated because of people's fears of contracting it by touch. As people learned that HIV was transmitted only through bodily fluids, many people's initial fears were reduced, and patients were no longer stigmatized. Beware, however, of excessive familiarity that may breed complacency. If something is genuinely risky, it pays to keep your guard up.

7. Evolution works at the level of the individual. We should recognize that we are the product of natural selection when we pay attention to our personal welfare, the welfare of our kin, and perhaps that of close associates who aren't kin. As we learned in Chapter 8, ground squirrels are quite sensitive to their audience when emitting risky alarm calls and are more likely to call when close relatives are within earshot. Squirrels who adopt this strategy leave more descendants than those who emit calls with less discrimination.

We too are primed to care about relatives and individuals more than population-level statistics. Most gun violence incidents have a single victim. For many of us, these victims are strangers. Media reporting that personalizes these murders and describes the suffering of their survivors may be an effective way to communicate the tragedy of gun violence and make the risks and costs more tangible. Following a terrorist attack or other mass casualty event, reports that personalize the victims are those that are most likely to help us understand the magnitude of the event. Politicians know this and often try to manipulate our perceptions of security by telling stories about individual victims, which they link to a larger problem. Their policy (border walls, health care reform, and so on) is then proposed as a solution. I saw this firsthand when Los Angeles mayor Eric Garcetti shuttered the city and instructed Angelenos to stay at home during the COVID-19 pandemic. Social distancing was presented as a measure to save our neighbors and our relatives and to reduce the burden on our health-care providers, who were desperately working to save lives. If we understand why we respond to messages that tell stories about individuals, we can then arm ourselves with data and do a better job of objectively evaluating the true risk.

8. The scope of damage may also influence our evaluation of risks. Is damage concentrated on an individual or spread across the environment? Most organisms occupy relatively small patches of the Earth and have evolved mechanisms to make assessments about things that happen in areas they occupy. Marmots are more vulnerable when they are far from protective burrows, and they perceive greater risks in areas with limited peripheral visibility. Bushbuck, however, seek dense wooded areas for safety from large carnivores. If these species' local landscapes are modified, their risk assessments will also change.

Yet, the scope of damage that we've created is global. It's extremely difficult for us to link our personal carbon footprint to the consequences of the sum total of our global energy consumption and release of carbon dioxide. We can't comprehend the spatial scale of a nuclear war, which has a global impact. And, we can't understand the temporal scale of a nuclear meltdown. A really bad meltdown in a populated area could easily kill or shorten the lives of numerous people while making the area uninhabitable for many generations. By personalizing the damage and by putting a human face on it, we may be able to better understand the risks. Sadly, climate-driven catastrophes are becoming much more common, and we can see all too well the suffering that follows a superstorm. We evolved from a long line of storytellers, and good stories can effectively spur us to action. To ensure that it's the right action, consider verifying the sources of information.

9. Fear can depend on location. Elk in the northern part of Yellowstone National Park maintain a complex and temporally variable assessment of where they are likely to encounter wolves. Much like the elk, it's to be expected that we may become anxious at certain locations where we've had bad experiences. Stepping back a bit, this insight can explain some odd human reactions. Janice, my wife, was driving her visiting parents to the beach one day when a red-light camera flashed. She later recounted to me that she was crossing Beethoven Street in Culver City, California, and thought that the yellow light turned red while she was still in the intersection. She dreaded the day when the automatic traffic ticket would appear in the mailbox. Remarkably, it never did. But for at least ten years after this experience, whenever we approached Beethoven Street (and *only* Beethoven Street), she pointed out the red-light camera. To get to Beethoven Street, of course, we'd probably driven through ten other red-light cameras that didn't bother her. By understanding our inner marmot, we can explain some of our own peculiar behaviors and perhaps those of our loved ones.

10. Risk is context dependent. We have learned that our physiological state can influence our perceptions of risk, as we saw with the decisions marmots make about whether to emit alarm calls. This means that we should expect our state or condition and a variety of other external stimuli to influence both our actual well-being and safety, and our perceptions of well-being and safety. In the aftermath of a failed 2016 coup in Turkey, there was a widespread purge of university faculty. For this reason tenured university professors in Turkey had much more to fear from speaking their mind than tenured professors at the University of California. The fear of punishment has major impacts on personal well-being and optimal responses in a given situation.

11. We systematically overestimate small risks and underestimate large risks. For instance, we overestimate the number of deaths due to botulism and underestimate the number of deaths due to cancer. Error management theory (EMT) explains the underlying biological basis of our biases. This theory predicts that we should behave in ways consistent with the folk advice of nothing-ventured-nothing-gained when the benefits are huge, even if the chance of success is small. EMT also reveals that if the cost of a mistake is particularly large, we should behave in conservative ways. In many situations, particularly when the cost of a mistake is great, overestimating risk is generally adaptive. We saw this with Randy Nesse's smoke detector principle (it's better to have your smoke alarm go off when you burn your toast to ensure that it will go off when there's a real fire), and error management theory gives us the tools to understand when and why. Our anxiety is adaptive, and having cautious responses makes sense when the cost of errors is especially high. A large empirical decision-making literature reveals specific biases. Despite all of the idiosyncrasies that make us human, formal decision theory, which mathematically calculates optimal decisions, provides a valuable framework with which we can evaluate risk.

Human decision theorists have come up with statistical methods to integrate different attributes of risk. One powerful statistical technique combines them in a way that allows us to visualize risks on two different dimensions. One dimension includes the following risk attributes: how involuntary or delayed an outcome is, how unknown it is, how well science understands it, how uncontrollable it is, how catastrophic it is, and how dreadful it is. Along this axis, risks like nuclear power, pesticides, and food coloring score particularly high while skiing, alcoholic beverages, swimming, and mountain climbing score particularly low. People make conscious decisions to ski, drink, swim, and climb mountains, but they are subjected to nuclear power, pesticides, and food coloring. A second dimension further differentiates these risks and is characterized by the degree of certainty about whether an exposure is fatal, how dreadful it is, and how catastrophic it is. Along this axis, risks like general aviation, handguns, and nuclear power score high, while home appliances, power mowers, and food coloring score low.

On these two dimensions, therefore, both guns and nuclear power are seen as certainly fatal and are perceived with some dread, but nuclear power is viewed as involuntary, and its effects are delayed. Thus, we expect different assessments about these risks, and this is why nuclear power is perceived as being more risky than guns. To properly use our fear to thrive, we must hone our risk assessment abilities. We must become comfortable making decisions under uncertainty, and we must wisely let context dictate the correct decisions. We should embrace decision theory and use it more when making consequential decisions—including about whom to vote for.

12. Sociality can buffer fear. Whether it's by grouping to avoid predators or just listening to others, animals receive antipredator benefits from being around each other. Recent work has even suggested that social buffering can block fear conditioning in rats. I suggest that this applies to us too. Fears and other emotions can be contagious, and by being sensitive to others' emotions we can quickly respond to threats. These responses are likely rooted in our capacity for empathy.

But these days empathy is in short supply. We are quick to point the finger at those who have different beliefs without trying to understand where they are coming from. We harden our positions, associate with those who share similar ones, and lose respect for others' positions. Worse yet, as Sherry Turkle argues in *Reclaiming Conversation,* many of us avoid conversations, preferring to communicate by text, which eliminates all the social cues so important for meaningful, empathic communication. Text-based communication may be exacerbated by open-plan offices. Such office layouts are marketed as a structured way to encourage interactions, so it's a shame that the data suggest otherwise. When Ethan Bernstein and Stephen Turban quantified how office workers communicated in open-plan and more traditional offices, they found that people retreated to their own desks or cubicles in open-plan offices and actually communicated more by text than in person.

One way to reclaim empathy is to have more dinner parties with our neighbors to discuss and debate controversial topics and the fears that underlie them. By talking with and listening to others, we can understand why people have views divergent from our own. By having meaningful conversations about our fears with others we let our capacity for empathy reduce them. Of course, social transmission is a force multiplier that could go the other way as well. So we must guard against having our fears socially enhanced, and we must guard against only listening to people with similar views. To make the best decisions it is essential to seek out contradictory perspectives to challenge our assumptions. A growing literature shows that when groups composed of diverse people with different perspectives work together, they make better decisions than less diverse groups.

13. Learning, whether it is through fear conditioning or habituation, has a profound impact on our assessments of risk. We know that animals have some innate predispositions to respond fearfully to certain things, but as we learned from tammar wallabies and a variety of fishes, experience with predators is often required to hone those abilities. For complex, long-lived species like us, we should expect that learning is an important mechanism in how we respond fearfully to various events and objects. We have biases to learn to fear snakes and spiders even if we'll never really be threatened by them.

Our culture, which is driven by social learning, teaches us both reasonable and unreasonable fears. We should be aware of the force-multiplying effect of social learning and realize that we may sometimes learn the wrong things to fear. I believe this risk of learning the wrong things compels us to properly estimate the risks of events and base our decisions, whether personal or political, on the best evidence available.

14. We must be aware of our surroundings. Like the calling peacock or the go-away birds that communicate to their entire community that something scary is around, when in threatening situations we should be open to new sources of information to help us better estimate risk, even if it's untraditional. Yet we should constantly evaluate these sources of potential information. From an ecological perspective, the ubiquitous connectedness between different species means that maintaining diverse ecological communities may be essential for their survival, as well as for ours.

15. We are especially vulnerable to those who seek to manipulate us with fear. Manipulating fear in others can be an effective way to motivate change, in certain circumstances. Fear can be used to nudge us into healthier practices, get us to evacuate before a hurricane, and prepare us for earthquakes. Our loss-aversion bias can be used profitably to encourage people to make properly informed decisions about the true risks of a medicine, a surgical procedure, or the costs of things like pollution.

But this knowledge can also be used maliciously, and we should guard against becoming a target and protect others from becoming targets. One way to do so is to pause and reframe the question. For instance, rather than focusing on how many people are killed by terrorist attacks annually, focus on how many people are not killed by terrorist attacks. The probability of being killed by a terrorist attack, in most countries, is fortunately quite small. Another way to inoculate people against misinformation is to show how sensitive we are to misinformation. Given this sensitivity, we can teach people to be skeptical about simple messages designed to instill fear, and to demand evidence of a simple causal link.

But we have a new challenge. We are in the midst of a great mismatch that we have created by culturally modifying our environment, and the decision rules we've evolved may no longer be relevant. Like the birds that can't estimate the velocity of an approaching object above a certain threshold because they have not evolved mechanisms to avoid hitting an approaching airplane, we also have not evolved to comprehend many novel features of our modern environment. We used to fight with sticks and rocks and can't really estimate the impact of nuclear weapons. We have evolved to modify our behavior in response to immediate changes in temperature; we put on sweaters if it's cold and remove our hands quickly from heat. But we are unable to properly respond by modifying our behavior now to prevent even worse anthropogenic climate change at some point in the future. Yet, if our behavioral changes caused the problems, our behavioral changes must also be the solution.

We do have the power to make certain thoughtful behavioral changes in response to stimuli we're evolutionarily unprepared to deal with. One example of such a stimulus is the twenty-four-hour updates we are constantly receiving of threatening news, available at all times on our television sets, cell phones, and computers. We may habituate to this overstimulation, but if so, we risk becoming less sensitive to important messages. Losing our fear removes something vital to who we are—we become less human. We've evolved to respond immediately to threats by engaging a sophisticated series of neurochemical responses that have served us and our ancestors well in the past. Yet for many of us, the constant activation of our HPA axis just causes stress and a feeling of indecision. We should be aware of these responses and, if necessary, reduce the rate at which we receive information so that we can respond appropriately when necessary.

As I age, I become more cautious. And, at a very proximate level, the structure of our brain changes as we age. A recent study found that the right parietal cortex shrinks with age and the relative size of this part of the brain is associated with risk taking. I don't think that I've become more fearful per se, but the consequences of decisions have changed since I was younger and stronger and more resilient. This is natural.

While I grew up in the ocean—swimming, body surfing, and body boarding—it wasn't until I was forty that I really began to surf. I've had precious moments with my family in the ocean—like the day when David and I sat in awe while two pairs of dolphins herded a huge mixed species school of fish into a bait ball beneath us. Brown pelicans and double-crested cormorants dove next to us into the feeding frenzy, emerging triumphantly with fish, only to be attacked by western gulls who, directly out of a scene in *Finding Nemo,* sought to claim the fish as theirs. Or the days when seals or sea lions poke their heads up next to us, fuzzy whiskers reflecting in the morning sun, curiously checking us out. Or the days when it is quiet, and the waves come slowly, and we just sit and enjoy the moment.

I know that I will never surf a big wave. I also will no longer ski down forty-five- to fifty-degree chutes after ascending them with ice axes and ropes. I will never climb that 7,000-meter peak or climb long, exposed rocky arêtes. Those days are all in my past. These days I'm quite content to hike to a pretty meadow or overlook, take a nap, and, upon waking, see the life and beauty in nature.

Now, with the responsibilities of family and home ownership, I have more to lose, and I crave more stability. Because I have something to lose, I have fear of losing whatever stability I perceive I have.

But I find it comforting to know that my fear comes from a long line of my ancestors, both human and nonhuman. It is an inherited treasure, a powerful ally. Yet, it is also an annoying and sometimes intolerable companion. It is a compass that, when calibrated properly, guides us away from danger and toward opportunity.

At some level, our relationship with fear is a lesson from life. Since it's impossible to eliminate risk, our fears and anxieties assist us in making the right decisions. Since we cannot eliminate them, we should both embrace our fears and challenge them. As Mary Schmich, a journalist at the *Chicago Tribune,* wrote in 1997, "Do one thing every day that scares you."

Further Reading

Knowledge is built on the shoulders of our intellectual ancestors, and much of what I know comes from reading other people's studies. However, this is not a dense academic tome, and thus to enhance readability, a strategic decision was made to limit the references presented to a handful of key sources, including in particular those readily available on the internet or through the library system.

Prologue

"Cognitive Bias Codex." Categorization by Buster Benson. Design by John Manoogian III. Available from Wikimedia Commons, https://commons .wikimedia.org/wiki/File:The_Cognitive_Bias_Codex_-_180%2B_biases, _designed_by_John_Manoogian_III_(jm3).png.

de Waal, Frans B. M. "Anthropomorphism and Anthropodenial: Consistency in Our Thinking about Humans and Other Animals." *Philosophical Topics* 27 (1999): 255–280.

DuPont, Robert L., Dorothy P. Rice, Leonard S. Miller, Sarah S. Shiraki, Clayton R. Rowland, and Henrick J. Harwood. "Economic Costs of Anxiety Disorders." *Anxiety* 2 (1996): 167–172.

Lépine, Jean-Pierre. "The Epidemiology of Anxiety Disorders: Prevalence and Societal Costs." *Journal of Clinical Psychiatry* 14 (2002): 4–8.

Natterson-Horowitz, Barbara, and Kathryn Bowers. *Zoobiquity: The Astonishing Connection between Human and Animal Health.* New York: Vintage, 2013.

Pimm, Stuart L., Clinton N. Jenkins, Robin Abell, Thomas M. Brooks, John L. Gittleman, Lucas N. Joppa, Peter H. Raven, Callum M. Roberts, and Joseph O. Sexton. "The Biodiversity of Species and Their Rates of Extinction, Distribution, and Protection." *Science* 344 (2014): 1246752.

Shubin, Neil. *Your Inner Fish: A Journey into the 3.5-Billion-Year History of the Human Body.* New York: Vintage, 2008.

1. A Sophisticated Neurochemical Cocktail

Balavoine, Guillaume, and André Adoutte. "The Segmented Urbilateria: A Testable Scenario." *Integrative and Comparative Biology* 43 (2003): 137–147.

Bercovitch, Fred B., Marc D. Hauser, and James H. Jones. "The Endocrine Stress Response and Alarm Vocalizations in Rhesus Macaques." *Animal Behaviour* 49 (1995): 1703–1706.

Blumstein, Daniel T., Janet Buckner, Sajan Shah, Shane Patel, Michael E. Alfaro, and Barbara Natterson-Horowitz. "The Evolution of Capture Myopathy in Hooved Mammals: A Model for Human Stress Cardiomyopathy?" *Evolution, Medicine, and Public Health* 2015 (2015): 195–203.

Blumstein, Daniel T., Benjamin Geffroy, Diogo S. M. Samia, and Eduardo Bessa, eds. *Ecotourism's Promise and Peril: A Biological Evaluation.* Cham, Switzerland: Springer, 2017.

Blumstein, Daniel T., Marilyn L. Patton, and Wendy Saltzman. "Faecal Glucocorticoid Metabolites and Alarm Calling in Free-Living Yellow-Bellied Marmots." *Biology Letters* 2 (2006): 29–32.

Goymann, Wolfgang, and John C. Wingfield. "Allostatic Load, Social Status and Stress Hormones: The Costs of Social Status Matter." *Animal Behaviour* 67 (2004): 591–602.

McNaughton, Neil, and Philip J. Corr. "A Two-Dimensional Neuropsychology of Defense: Fear / Anxiety and Defensive Distance." *Neuroscience and Biobehavioral Reviews* 28 (2004): 285–305.

Mobbs, Dean, and Jeansok J. Kim. "Neuroethological Studies of Fear, Anxiety, and Risky Decision-Making in Rodents and Humans." *Current Opinion in Behavioral Sciences* 5 (2015): 8–15.

Mobbs, Dean, Predrag Petrovic, Jennifer L. Marchant, Demis Hassabis, Nikolaus Weiskopf, Ben Seymour, Raymond J. Dolan, and Christopher D. Frith. "When Fear Is Near: Threat Imminence Elicits Prefrontal-Periaqueductal Gray Shifts in Humans." *Science* 317 (2007): 1079–1083.

Natterson-Horowitz, Barbara, and Kathryn Bowers. *Zoobiquity: The Astonishing Connection between Human and Animal Health.* New York: Vintage, 2013.

Nesse, Randolph M., and Elizabeth A. Young. "Evolutionary Origins and Functions of the Stress Response." In *Encyclopedia of Stress,* 3 vols., ed. George Fink, 2: 79–84. San Diego: Academic Press, 2000.

Nesse, R. M., S. Bhatnagar, and B. Ellis. "Evolutionary Origins and Functions of the Stress Response System." In *Stress: Concepts, Cognition, Emotion, and Behavior,* ed. George Fink, 95–101. Handbook of Stress, vol. 1. London: Academic Press, 2016.

Nilsson, Stefan. "Comparative Anatomy of the Autonomic Nervous System." *Autonomic Neuroscience: Basic and Clinical* 165 (2011): 3–9.

Sheriff, Michael J., Charles J. Krebs, and Rudy Boonstra. "The Ghosts of Predators Past: Population Cycles and the Role of Maternal Programming under Fluctuating Predation Risk." *Ecology* 91 (2010): 2983–2994.

———. "The Sensitive Hare: Sublethal Effects of Predator Stress on Reproduction in Snowshoe Hares." *Journal of Animal Ecology* 78 (2009): 1249–1258.

Wingfield, John C., Donna L. Maney, Creagh W. Breuner, Jerry D. Jacobs, Sharon Lynn, Marilyn Ramenofsky, and Ralph D. Richardson. "Ecological Bases of Hormone-Behavior Interactions: The 'Emergency Life History Stage.'" *American Zoologist* 38 (1998): 191–206.

2. Beware of Looming Objects

Blumstein, Daniel T. "Moving to Suburbia: Ontogenetic and Evolutionary Consequences of Life on Predator-Free Islands." *Journal of Biogeography* 29 (2002): 685–692.

———. "The Multipredator Hypothesis and the Evolutionary Persistence of Antipredator Behavior." *Ethology* 112 (2006): 209–217.

Blumstein, Daniel T., Janice C. Daniel, Andrea S. Griffin, and Christopher S. Evans. "Insular Tammar Wallabies (*Macropus eugenii*) Respond to Visual but Not Acoustic Cues from Predators." *Behavioral Ecology* 11 (2000): 528–535.

Blumstein, Daniel T., Janice C. Daniel, and Brian P. Springett. "A Test of the Multi-Predator Hypothesis: Rapid Loss of Antipredator Behavior after 130 Years of Isolation." *Ethology* 110 (2004): 919–934.

Burger, Joanna, Michael Gochefeld, and Bertram G. Murray Jr. "Role of a Predator's Eye Size in Risk Perception by Basking Black Iguana, *Ctenosaura similis.*" *Animal Behaviour* 42 (1991): 471–476.

Chan, Alvin Aaden Yim-Hol, Paulina Giraldo-Perez, Sonja Smith, and Daniel T. Blumstein. "Anthropogenic Noise Affects Risk Assessment and Attention: The Distracted Prey Hypothesis." *Biology Letters* 6 (2010): 458–461.

Chan, Alvin Aaden Yim-Hol, W. David Stahlman, Dennis Garlick, Cynthia D. Fast, Daniel T. Blumstein, and Aaron P. Blaisdell. "Increased Amplitude and Duration of Acoustic Stimuli Enhance Distraction." *Animal Behaviour* 80 (2010): 1075–1079.

Cook, Michael, and Susan Mineka. "Selective Associations in the Observational Conditioning of Fear in Rhesus Monkeys." *Journal of Experimental Psychology Animal Behavior Processes* 16 (1990): 372–389.

Curio, Eberhard. "The Functional Organization of Anti-Predator Behaviour in the Pied Flycatcher: A Study of Avian Visual Perception." *Animal Behaviour* 23 (1975): 1–115.

DeVault, Travis L., Bradley F. Blackwell, Thomas W. Seamans, Steven L. Lima, and Esteban Fernández-Juricic. "Effects of Vehicle Speed on Flight Initiation by Turkey Vultures: Implications for Bird-Vehicle Collisions." *PLoS One* 9 (2014): e87944.

———. "Speed Kills: Ineffective Avian Escape Responses to Oncoming Vehicles." *Proceedings of the Royal Society B* 282 (2015): 20142188.

Griffin, Andrea S., Christopher S. Evans, and Daniel T. Blumstein. "Learning Specificity in Acquired Predator Recognition." *Animal Behaviour* 62 (2001): 577–589.

———. "Selective Learning in a Marsupial." *Ethology* 108 (2002): 1103–1014.

Kawai, Nobuyuki, and Hongshen He. "Breaking Snake Camouflage: Humans Detect Snakes More Accurately Than Other Animals under Less Discernible Visual Conditions." *PLoS One* 11 (2016): e0164342.

Lima, Steven L., Bradley F. Blackwell, Travis L. DeVault, and Esteban Fernández-Juricic. "Animal Reactions to Oncoming Vehicles: A Conceptual Review." *Biological Reviews of the Cambridge Philosophical Society* 90 (2015): 60–76.

Mobbs, Dean, Rongjun Yu, James B. Rowe, Hannah Eich, Oriel FeldmanHall, and Tim Dalgleish. "Neural Activity Associated with Monitoring the Oscillating Threat Value of a Tarantula." *Proceedings of the National Academy of Science USA* 107 (2010): 20582–20586.

Rakison, David H., and Jaime Derringer. "Do Infants Possess an Evolved Spider-Detection Mechanism? *Cognition* 107, no. 1 (2008): 381–393.

Shibasaki, Masahiro, and Nobuyuki Kawai. "Rapid Detection of Snakes by Japanese Monkeys (*Macaca fuscata*): An Evolutionarily Predisposed Visual System." *Journal of Comparative Psychology* 123 (2009): 131–135.

Van Le, Quan, Lynne A. Isbell, Jumpei Matsumoto, Minh Nguyen, Etsuro Hori, Rafael S. Maior, Carlos Tomaz, et al. "Pulvinar Neurons Reveal Neuro-biological Evidence of Past Selection for Rapid Detection of Snakes." *Proceedings of the National Academy of Science USA* 110 (2013): 19000–19005.

Yorzinski, Jessica L., Michael J. Penkunas, Michael L. Platt, and Richard G. Coss. "Dangerous Animals Capture and Maintain Attention in Humans." *Evolutionary Psychology* 12 (2014): 534–548.

3. Noise Matters

Arnal, Luc H., Adeen Flinker, Andreas Kleinschmidt, Anne-Lise Giraud, and David Poeppel. "Human Screams Occupy a Privileged Niche in the Communication Soundscape." *Current Biology* 25 (2015): 2051–2056.

Bledsoe, Ellen K., and Daniel T. Blumstein. "What Is the Sound of Fear? Behavioral Responses of White-Crowned Sparrows *Zonotrichia leucophrys* to Synthesized Nonlinear Acoustic Phenomena." *Current Zoology* 60 (2014): 534–541.

Blumstein, Daniel T., Greg A. Bryant, and Peter Kaye. "The Sound of Arousal in Music Is Context-Dependent." *Biology Letters* 8 (2012): 744–747.

Blumstein, Daniel T., Louise Cooley, Jamie Winternitz, and Janice C. Daniel. "Do Yellow-Bellied Marmots Respond to Predator Vocalizations?" *Behavioral Ecology and Sociobiology* 62 (2008): 457–468.

Blumstein, Daniel T., Richard Davitian, and Peter D. Kaye. "Do Film Soundtracks Contain Nonlinear Analogues to Influence Emotion?" *Biology Letters* 6 (2010): 751–754.

Blumstein, Daniel T., and Charlotte Recapet. "The Sound of Arousal: The Addition of Novel Non-Linearities Increases Responsiveness in Marmot Alarm Calls." *Ethology* 115 (2009): 1074–1081.

Blumstein, Daniel T., Dominique T. Richardson, Louise Cooley, Jamie Winternitz, and Janice C. Daniel. "The Structure, Meaning and Function of Yellow-Bellied Marmot Pup Screams." *Animal Behaviour* 76 (2008): 1055–1064.

"Children, Youth, Families and Socioeconomic Status." Fact Sheet, American Psychological Association, n.d.. https://www.apa.org/pi/ses/resources /publications/factsheet-cyf.pdf.

Coleman, Andrea, Dominique Richardson, Robin Schechter, and Daniel T. Blumstein. "Does Habituation to Humans Influence Predator Discrimination in Gunther's Dik-Diks (*Madoqua guentheri*)?" *Biology Letters* 4 (2008): 250–252.

Darwin, Charles. *The Expression of Emotions in Man and Animals.* London: John Murray, 1872.

Ekman, Paul. "Facial Expression and Emotion." *American Psychologist* 48 (1993): 384–392.

Hettena, Alexandra M., Nicole Munoz, and Daniel T. Blumstein. "Prey Responses to Predator's Sounds: A Review and Empirical Study." *Ethology* 120 (2014): 427–452.

Johnson, Frances R., Elisabeth J. McNaughton, Courtney D. Shelley, and Daniel T. Blumstein. "Mechanisms of Heterospecific Recognition in Avian Mobbing Calls." *Australian Journal of Zoology* 51 (2003): 577–585.

Lea, Amanda J., June P. Barrera, Lauren M. Tom, and Daniel T. Blumstein. "Heterospecific Eavesdropping in a Nonsocial Species." *Behavioral Ecology* 19 (2008): 1041–1046.

McEwen, Bruce S. "Effects of Stress on the Developing Brain." *Cerebrum* 2011 (2011): 14.

Slaughter, Emily I., Erin R. Berlin, Jonathan T. Bower, and Daniel T. Blumstein. "A Test of the Nonlinearity Hypothesis in Great-Tailed Grackles (*Quiscalus mexicanus*)." *Ethology* 119 (2013): 309–315.

Zanette, Liana Y., Aija F. White, Marek C. Allen, and Michael Clinchy. "Perceived Predation Risk Reduces the Number of Offspring Songbirds Produce Per Year." *Science* 334 (2011): 1398–1401.

4. Smells Risky to Me

Apfelbach, Raimund C., Dixie Blanchard, Robert J. Blanchard, Richard A. Hayes, and Iain S. McGregor. "The Effects of Predator Odors in Mammalian Prey Species: A Review of Field and Laboratory Studies." *Neuroscience and Biobehavioral Reviews* 29, no. 8 (2005): 1123–1144.

Arshamian, Artin, Matthias Laska, Amy R. Gordon, Matilda Norberg, Christian Lahger, Danja K. Porada, Nadia Jelvez Serra, et al. "A Mammalian Blood Odor Component Serves as an Approach-Avoidance Cue across Phylum Border—from Flies to Humans." *Scientific Reports* 7 (2017): 13635.

Berdoy, M., J. P. Webster, and D. W. Macdonald. "Fatal Attraction in Rats Infected with *Toxoplasma gondii*." *Proceedings of the Royal Society B* 267 (2000): 1591–1594.

Blumstein, Daniel T., Lisa Barrow, and Markael Luterra. "Olfactory Predator Discrimination in Yellow-Bellied Marmots." *Ethology* 114 (2008): 1135–1143.

Dewan, Adam, Rodrigo Pacifico, Ross Zhan, Dmitry Rinberg, and Thomas Bozza. "Non-Redundant Coding of Aversive Odours in the Main Olfactory Pathway." *Nature* 497 (2013): 486–489.

Ferrari, Maud C. O., Brian D. Wisenden, and Douglas P. Chivers. "Chemical Ecology of Predator–Prey Interactions in Aquatic Ecosystems: A Review and Prospectus." *Canadian Journal of Zoology* 88 (2010): 698–724.

Ferrero, David M., Jamie K. Lemon, Daniela Fluegge, Stan L. Pashkovski, Wayne J. Korzan, Sandeep R. Datta, Marc Spehr, Markus Fendt, and Stephen D. Liberles. "Detection and Avoidance of a Carnivore Odor by Prey." *Proceedings of the National Academy of Science USA* 108 (2011): 11235–11240.

Fessler, Daniel, and Kevin Haley. "Guarding the Perimeter: The Outside-inside Dichotomy in Disgust and Bodily Experience." *Cognition and Emotion* 20 (2006): 3–19.

Flegr, J. "Influence of Latent Toxoplasma Infection on Human Personality, Physiology and Morphology: Pros and Cons of the Toxoplasma-Human Model in Studying the Manipulation Hypothesis." *Journal of Experimental Biology* 216 (2013): 127–133.

Johnson, Stefanie K., Markus A. Fitza, Daniel A. Lerner, Dana M. Calhoun, Marissa A. Beldon, Elsa T. Chan, and Pieter T. J. Johnson. "Risky Business: Linking *Toxoplasma gondii* Infection and Entrepreneurship Behaviours across Individuals and Countries." *Proceedings of the Royal Society B* 285 (2018): 20180822.

Jones, Menna E., Raimund Apfelbach, Peter B. Banks, Elissa Z. Cameron, Chris R. Dickman, Anke Frank, Stuart McLean, et al. "A Nose for Death: Integrating Trophic and Informational Networks for Conservation and Management." *Frontiers in Ecology and Evolution* 4 (2016): 124.

Lazenby, Bill T., and Christopher R. Dickman. "Patterns of Detection and Capture Are Associated with Cohabiting Predators and Prey." *PLoS One* 8 (2013): e59846.

McGann, John P. "Poor Human Olfaction Is a 19th-Century Myth." *Science* 356 (2017): eaam7263.

Parsons, Michael H., Raimund Apfelbach, Peter B. Banks, Elissa Z. Cameron, Chris R. Dickman, Anke S. K. Frank, Menna E. Jones, et al. "Biologically Meaningful Scents: A Framework for Understanding Predator-Prey Research across Disciplines." *Biological Reviews of the Cambridge Philosophical Society* 93 (2018): 98–114.

Parsons, Michael H., and Daniel T. Blumstein. "Familiarity Breeds Contempt: Kangaroos Persistently Avoid Areas with Experimentally Deployed Dingo Scents." *PLoS One* 5 (2010): e10403.

Parsons, Michael H., and Daniel T. Blumstein. "Feeling Vulnerable? Indirect Risk Cues Differently Influence How Two Marsupials Respond to Novel Dingo Urine." *Ethology* 116 (2010): 972–980.

Swihart, Robert K., Mary Jane I. Mattina, and Joseph J. Pignatello. "Repellency of Predator Urine to Woodchucks and Meadow Voles." National Wildlife

Research Center Repellents Conference, Denver, August 1995. *Proceedings of the Second DWRC Special Symposium,* ed. J. Russell Mason, 271–284.

Wisenden, Brian D. "Chemical Cues That Indicate Risk of Predation." In *Fish Pheromones and Related Cues,* ed. Peter W. Sorenson and Brian D. Wisendon, 131–148: Chichester, UK: John Wiley and Sons, 2015.

5. Be Very Aware

Bednekoff, Peter A., and Daniel T. Blumstein. "Peripheral Obstructions Influence Marmot Vigilance: Integrating Observational and Experimental Results." *Behavioral Ecology* 20 (2009): 1111–1117.

Berger, Joel, Jon E. Swenson, and Inga-Lill Persson. "Recolonizing Carnivores and Naive Prey: Conservation Lessons from Pleistocene Extinctions." *Science* 291 (2001): 1036–1039.

Blumstein, Daniel T. "Quantifying Predation Risk for Refuging Animals: A Case Study with Golden Marmots." *Ethology* 104 (1998): 501–516.

Blumstein, Daniel T., and Janice C. Daniel. "Isolation from Mammalian Predators Differentially Affects Two Congeners." *Behavioral Ecology* 13 (2002): 657–663.

Brown, Joel S. "Vigilance, Patch Use and Habitat Selection: Foraging under Predation Risk." *Evolutionary Ecology Research* 1 (1999): 49–71.

Clearwater, Yvonne A., and Richard G. Coss. "Functional Esthetics to Enhance Well-Being in Isolated and Confined Settings." In *From Antarctica to Outer Space,* ed. A. A. Harrison, Y. A. Clearwater, and C. P. McKay, 331–348. New York: Springer, 1991.

Coss, Richard G., and Eric P. Charles. "The Role of Evolutionary Hypotheses in Psychological Research: Instincts, Affordances, and Relic Sex Differences." *Ecological Psychology* 16 (2004): 199–236.

Coss, Richard G., and Ronald O. Goldthwaite. "The Persistence of Old Designs for Perception." *Perspectives in Ethology* 11 (1995): 83–148.

Coss, Richard G., and Michael Moore. "Precocious Knowledge of Trees as Antipredator Refuge in Preschool Children: An Examination of Aesthetics, Attributive Judgments, and Relic Sexual Dinichism." *Ecological Psychology* 14 (2002): 181–222.

Dukas, Reuven, and Alan C. Kamil. "The Cost of Limited Attention in Blue Jays." *Behavioral Ecology* 11 (2000): 502–506.

Ely, Craig R., David H. Ward, and Karen S. Bollinger. "Behavioral Correlates of Heart Rates of Free-Living Greater White-Fronted Geese." *Condor* 101 (1999): 390–395.

Herodotus. *The History of Herodotus*. Translated by George Rawlinson. 3rd ed. New York: Scribner, 1875.

Lankston, Louise, Pearce Cusack, Chris Fremantle, and Chris Isles. "Visual Art in Hospitals: Case Studies and Review of the Evidence." *Journal of the Royal Society of Medicine* 103 (2010): 490–499.

Leiner, Lisa, and Markus Fendt. "Behavioural Fear and Heart Rate Responses of Horses after Exposure to Novel Objects: Effects of Habituation." *Applied Animal Behaviour Science* 131 (2011): 104–109.

Monclús, Raquel, Alexandra M. Anderson, and Daniel T. Blumstein. "Do Yellow-Bellied Marmots Perceive Enhanced Predation Risk When They Are Farther from Safety? An Experimental Study." *Ethology* 121 (2015): 831–839.

Orians, Gordon H. *Snakes, Sunrises, and Shakespeare: How Evolution Shapes Our Loves and Fears.* Chicago: University of Chicago Press, 2014.

Piper, Walter H. "Exposure to Predators and Access to Food in Wintering White-Throated Sparrows *Zonotrichia albicollis*." *Behaviour* 112 (1990): 284–298.

Prins, H. H. T., and G. R. Iason. "Dangerous Lions and Nonchalant Buffalo." *Behaviour* 108 (1989): 262–296.

Simons, Marlise. "Himalayas Offer Clue to Legend of Gold-Digging 'Ants.'" *New York Times,* November 25, 1996, A5.

6. Economic Logic

Blumstein, Daniel T. "Developing an Evolutionary Ecology of Fear: How Life History and Natural History Traits Affect Disturbance Tolerance in Birds." *Animal Behaviour* 71 (2006): 389–399.

———. *An Ecotourist's Guide to Khunjerab National Park.* Lahore: World Wide Fund for Nature-Pakistan, 1995.

———. "Flush Early and Avoid the Rush: A General Rule of Antipredator Behavior?" *Behavioral Ecology* 21 (2010): 440–442.

Blumstein, Daniel T., Esteban Fernández-Juricic, Patrick A. Zollner, and Susan C. Garity. "Inter-Specific Variation in Avian Responses to Human Disturbance." *Journal of Applied Ecology* 42 (2005): 943–953.

Blumstein, Daniel, Benjamin Geffroy, Diogo Samia, and Eduardo Bessa. "Ecotourism Could Be Making Animals Less Scared, and Easier to Eat." *The Conversation* website, October 22, 2015. https://theconversation.com /ecotourism-could-be-making-animals-less-scared-and-easier-to-eat -49196.

Blumstein, Daniel T., Benjamin Geffroy, Diogo S. M. Samia, and Eduardo Bessa, eds. *Ecotourism's Promise and Peril: A Biological Evaluation*. Cham, Switzerland: Springer, 2017.

Cooper, William E., and Daniel T. Blumstein, eds. *Escaping from Predators: An Integrative View of Escape Decisions*. New York: Cambridge University Press, 2015.

Darwin, Charles. *Journal of Researches into the Geology and Natural History of the Various Countries Visited by HMS Beagle, under the Command of Captain Fitzroy from 1832 to 1836*. London: Colburn, 1840.

Geffroy, Benjamin, Diogo S. M. Samia, Eduardo Bessa, and Daniel T. Blumstein. "How Nature-Based Tourism Might Increase Prey Vulnerability to Predators." *Trends in Ecology and Evolution* 30 (2015): 755–765.

Samia, Diogo S. M., Daniel T. Blumstein, Mario Díaz, Tomas Grim, Juan Diego Ibáñez-Álamo, Jukka Jokimäki, Kunter Tätte, et al. "Rural-Urban Differences in Escape Behavior of European Birds across a Latitudinal Gradient." *Frontiers in Ecology and Evolution* 5 (2017): 66.

Samia, Diogo S. M., Daniel T. Blumstein, Theodore Stankowich, and William E. Cooper Jr. "Fifty Years of Chasing Lizards: New Insights Advance Optimal Escape Theory." *Biological Reviews of the Cambridge Philosophical Society* 91 (2016): 349–366.

Samia, Diogo S. M., Shinichi Nakagawa, Fausto Nomura, Thiago F. Rangel, and Daniel T. Blumstein. "Increased Tolerance to Humans among Disturbed Wildlife." *Nature Communications* 6 (2015): 8877.

Shaw, Mary, Richard Mitchell, and Danny Dorling. "Time for a Smoke? One Cigarette Reduces Your Life by 11 Minutes." *British Medical Journal* 320 (2000): 53.

Ydenberg, Ron C., and Lawrence M. Dill. "The Economics of Fleeing from Predators." *Advances in the Study of Behavior* 16 (1986): 229–249.

7. Once Bitten, Twice Shy

Blumstein, Daniel T. "Attention, Habituation, and Antipredator Behaviour: Implications for Urban Birds." In *Avian Urban Ecology*, ed. Diego Gil and Henrik Blum, 41–53. Oxford: Oxford University Press, 2014.

———. "Habituation and Sensitization: New Thoughts about Old Ideas." *Animal Behaviour* 120 (2016): 255–262.

Chivers, Douglas P., Mark I. McCormick, Matthew D. Mitchell, Ryan A. Ramasamy, and Maud C. O. Ferrari. "Background Level of Risk Determines How Prey Categorize Predators and Non-Predators." *Proceedings of the Royal Society B* 281 (2014): 20140355.

Cohen, Kristina L., Marc A. Seid, and Karen M. Warkentin. "How Embryos Escape from Danger: The Mechanism of Rapid, Plastic Hatching in Red-Eyed Treefrogs." *Journal of Experimental Biology* 219 (2016): 1875–1883.

Fazio, Lisa K., Nadia M. Brashier, B. Keith Payne, and Elizabeth J. Marsh. "Knowledge Does Not Protect against Illusory Truth." *Journal of Experimental Psychology: General* 144 (2015): 993–1002.

Ferrari, Maud C. O., François Messier, and Douglas P. Chivers. "Can Prey Exhibit Threat-Sensitive Generalization of Predator Recognition? Extending the Predator Recognition Continuum Hypothesis." *Proceedings of the Royal Society B* 275 (2008): 1811–1816.

Griffin, Andrea S., and Christopher S. Evans. "Social Learning of Antipredator Behaviour in a Marsupial." *Animal Behaviour* 66 (2003): 485–492.

Jeanty, Diane. "Rep. Duncan Hunter Now Fearmongering about Ebola as Well as Isis." *Huffington Post,* October 16, 2014. https://www.huffingtonpost.com /2014/10/16/duncan-hunter-isis-ebola_n_5997754.html.

Khalaf, Ossama, Siegfried Resch, Lucie Dixsaut, Victoire Gorden, Liliane Glauser, and Johannes Gräff. "Reactivation of Recall-Induced Neurons Contributes to Remote Fear Memory Attenuation." *Science* 360 (2018): 1239–1242.

King, Lucy E., Iain Douglas-Hamilton, and Fritz Vollrath. "African Elephants Run from the Sound of Disturbed Bees." *Current Biology* 17 (2007): R832–R833.

Perusini, Jennifer N., Edward M. Meyer, Virginia A. Long, Vinuta Rau, Nathaniel Nocera, Jacob Avershal, James Maksymetz, Igor Spigelman, and Michael S. Fanselow. "Induction and Expression of Fear Sensitization Caused by Acute Traumatic Stress." *Neuropsychopharmacology* 41 (2016): 45–57.

"Prolonged Exposure for PTSD." National Center for PTSD, U.S. Department of Veterans Affairs. https://www.ptsd.va.gov/understand_tx/prolonged _exposure.asp.

Rau, Vinuta, and Michael S. Fanselow. "Exposure to a Stressor Produces a Long Lasting Enhancement of Fear Learning in Rats: Original Research Report." *Stress* 12 (2009): 125–133.

Sebastian, Simone. "Examining 1962's 'Laughter Epidemic.'" *Chicago Tribune,* July 29, 2003.

Steketee, Jeffery D., and Peter W. Kalivas. "Drug Wanting: Behavioral Sensitization and Relapse to Drug-Seeking Behavior." *Pharmacological Reviews* 63 (2011): 348–365.

"World Urbanization Prospects, 2018 Revision." Population Division, United Nations Department of Economic and Social Affairs, May 16, 2018. https://www.un.org/development/desa/publications/2018-revision-of -world-urbanization-prospects.html.

8. Listening to Signalers

Barrera, June P., Leon Chong, Kaitlin N. Judy, and Daniel T. Blumstein. "Reliability of Public Information: Predators Provide More Information about Risk Than Conspecifics." *Animal Behaviour* 81 (2011): 779–787.

Benyus, Janine M. *Biomimicry: Innovation Inspired by Nature.* New York: Morrow, 1997.

Bercovitch, Fred B., Marc D. Hauser, and James H. Jones. "The Endocrine Stress Response and Alarm Vocalizations in Rhesus Macaques." *Animal Behaviour* 49 (1995): 1703–1706.

Blumstein, Daniel T. "The Evolution, Function, and Meaning of Marmot Alarm Communication." *Advances in the Study of Behavior* 37 (2007): 371–401.

Blumstein, Daniel T., Holly Fuong, and Elizabeth Palmer. "Social Security: Social Relationship Strength and Connectedness Influence How Marmots Respond to Alarm Calls." *Behavioral Ecology and Sociobiology* 71 (2017): 145.

Blumstein, Daniel T., and Olivier Munos. "Individual, Age and Sex-Specific Information Is Contained in Yellow-Bellied Marmot Alarm Calls." *Animal Behaviour* 69 (2005): 353–361.

Blumstein, Daniel T., Marilyn L. Patton, and Wendy Saltzman. "Faecal Glucocorticoid Metabolites and Alarm Calling in Free-Living Yellow-Bellied Marmots." *Biology Letters* 2 (2006): 29–32.

Blumstein, Daniel T., Jeff Steinmetz, Kenneth B. Armitage, and Janice C. Daniel. "Alarm Calling in Yellow-Bellied Marmots: II. The Importance of Direct Fitness." *Animal Behaviour* 53 (1997): 173–184.

Blumstein, Daniel T., Laure Verneyre, and Janice C. Daniel. "Reliability and the Adaptive Utility of Discrimination among Alarm Callers." *Proceedings of the Royal Society B:* 271 (2004): 1851–1857.

Carrasco, Malle F., and Daniel T. Blumstein. "Mule Deer (*Odocoileus hemionus*) Respond to Yellow-Bellied Marmot (*Marmota flaviventris*) Alarm Calls." *Ethology* 118 (2012): 243–250.

Cheney, Dorothy L., and Robert M. Seyfarth. *How Monkeys See the World: Inside the Mind of Another Species.* Chicago: University of Chicago Press, 1990.

Clay, Zanna, Carolynn L. Smith, and Daniel T. Blumstein. "Food-Associated Vocalizations in Mammals and Birds: What Do These Calls Really Mean?" *Animal Behaviour* 83 (2012): 323–330.

Ducheminsky, Nicholas, S. Peter Henzi, and Louise Barrett. "Responses of Vervet Monkeys in Large Troops to Terrestrial and Aerial Predator Alarm Calls." *Behavioral Ecology* 25 (2014): 1474–1484.

Evans, Christopher S. "Referential Communication." *Perspectives in Ethology* 12 (1997): 99–143.

Fuong, Holly, Adrianna Maldonado-Chaparro, and Daniel T. Blumstein. "Are Social Attributes Associated with Alarm Calling Propensity?" *Behavioral Ecology* 26 (2015): 587–592.

Goodale, Eben, Guy Beauchamp, and Graeme D. Ruxton. *Mixed-Species Groups of Animals: Behavior, Community Structure, and Conservation.* London: Academic Press, 2017.

Hingee, Mae, and Robert D. Magrath. "Flights of Fear: A Mechanical Wing Whistle Sounds the Alarm in a Flocking Bird." *Proceedings of the Royal Society B* 276 (2009): 4173–4179.

Lea, Amanda J., June P. Barrera, Lauren M. Tom, and Daniel T. Blumstein. "Heterospecific Eavesdropping in a Nonsocial Species." *Behavioral Ecology* 19 (2008): 1041–1046.

Magrath, Robert D., Tonya M. Haff, Jessica R. McLachlan, and Branislav Igic. "Wild Birds Learn to Eavesdrop on Heterospecific Alarm Calls." *Current Biology* 25 (2015): 2047–2050.

Manser, Marta B. "The Acoustic Structure of Suricates' Alarm Calls Varies with Predator Type and the Level of Response Urgency." *Proceedings of the Royal Society B* 268 (2001): 2315–2324.

Pollard, Kimberly A., and Daniel T. Blumstein. "Social Group Size Predicts the Evolution of Individuality." *Current Biology* 21 (2011): 413–417.

Preisser, Evan L., and John L. Orrock. "The Allometry of Fear: Interspecific Relationships between Body Size and Response to Predation Risk." *Ecosphere* 3 (2012): 1–27.

Shelley, Erin L., and Daniel T. Blumstein. "The Evolution of Vocal Alarm Communication in Rodents." *Behavioral Ecology* 16 (2004): 169–177.

Sherman, Paul W. "Nepotism and the Evolution of Alarm Calls." *Science* 197 (1977): 1246–1253.

Shriner, Walter McKee. "Yellow-Bellied Marmot and Golden-Mantled Ground Squirrel Responses to Heterospecific Alarm Calls." *Animal Behaviour* 55 (1998): 529–536.

Smith, Jennifer E., Raquel Monclús, Danielle Wantuck, Gregory L. Florant, and Daniel T. Blumstein. "Fecal Glucocorticoid Metabolites in Wild Yellow-Bellied Marmots: Experimental Validation, Individual Differences and Ecological Correlates." *General and Comparative Endocrinology* 178 (2012): 417–426.

9. Cascading Effects

Atkins, Justine L., Ryan A. Long, Johan Pansu, Joshua H. Daskin, Arjun B. Potter, Marc E. Stalmans, Corina E. Tarnita, and Robert M. Pringle. "Cascading Impacts of Large-Carnivore Extirpation in an African Ecosystem." *Science* 364 (2019): 173–177.

Darwin, Charles. *On the Origin of Species by Means of Natural Selection.* London: John Murray, 1859.

Kauffman, Matthew J., Jedediah F. Brodie, and Erik S. Jules. "Are Wolves Saving Yellowstone's Aspen? A Landscape-Level Test of a Behaviorally Mediated Trophic Cascade." *Ecology* 91 (2010): 2742–2755.

Kohl, Michel T., Daniel R. Stahler, Matthew C. Metz, James D. Forester, Matthew J. Kauffman, Nathan Varley, P. J. White, Douglas W. Smith, and Daniel R. MacNulty. "Diel Predator Activity Drives a Dynamic Landscape of Fear." *Ecological Monographs* 88 (2018): 638–652.

Lawrence, James, Katharina Schmid, and Miles Hewstone. 2019. "Ethnic Diversity, Ethnic Threat, and Social Cohesion: (Re)-Evaluating the Role of Perceived Out-Group Threat and Prejudice in the Relationship between Community Ethnic Diversity and Intra-Community Cohesion." *Journal of Ethnic and Migration Studies* 45 (2019): 395–418.

Letnic, Mike, and Freya Koch. "Are Dingoes a Trophic Regulator in Arid Australia? A Comparison of Mammal Communities on Either Side of the Dingo Fence." *Austral Ecology* 35 (2010): 167–175.

Moseby, Katherine E., Daniel T. Blumstein, and Mike Letnic. "Harnessing Natural Selection to Tackle the Problem of Prey Naïveté." *Evolutionary Applications* 9 (2016): 334–343.

Moseby, Katherine E., Amber Cameron, and Helen A. Crisp. "Can Predator Avoidance Training Improve Reintroduction Outcomes for the Greater Bilby in Arid Australia?" *Animal Behaviour* 83 (2012): 1011–1021.

Moseby, K. E., G. W. Lollback, and C. E. Lynch. "Too Much of a Good Thing; Successful Reintroduction Leads to Overpopulation in a Threatened Mammal." *Biological Conservation* 219 (2018): 78–88.

Putman, Robert D. *Bowling Alone: The Collapse and Revival of American Community.* New York: Simon and Schuster, 2000.

Ripple, William J., and Robert L. Beschta. "Trophic Cascades in Yellowstone: The First 15 Years after Wolf Reintroduction." *Biological Conservation* 145 (2012): 205–213.

Ripple, William J., James A. Estes, Oswald J. Schmitz, Vanessa Constant, Matthew J. Kaylor, Adam Lenz, Jennifer L. Motley, et al. "What Is a Trophic Cascade?" *Trends in Ecology and Evolution* 31 (2016): 842–849.

Sapolsky, Robert M. *Behave: The Biology of Humans at Our Best and Worst.* New York: Penguin, 2017.

Schmitz, Oswald J. "Behavior of Predators and Prey Links with Population Level Processes." In *Ecology of Predator-Prey Interactions,* ed. Pedro Barbosa and Ignacio Castellanos, 256–278. Oxford: Oxford University Press, 2005.

Suraci, Justin P., Michael Clinchy, Lawrence M. Dill, Devin Roberts, and Liana Y. Zanette. "Fear of Large Carnivores Causes a Trophic Cascade." *Nature Communications* 7 (2016): 10698.

Waser, Nickolas M., Mary V. Price, Daniel T. Blumstein, S. Reneé Arózqueta, Betsabé D. Castro Escobar, Richard Pickens, and Alessandra Pistoia. "Coyotes, Deer, and Wildflowers: Diverse Evidence Points to a Trophic Cascade." *Naturwissenschaften* 101 (2014): 427–436.

10. Minimizing Costs

Bouskila, Amos, and Daniel T. Blumstein. "Rules of Thumb for Predation Hazard Assessment: Predictions from a Dynamic Model." *American Naturalist* 139 (1992): 161–176.

Callaway, Ewen. "Genghis Khan's Genetic Legacy Has Competition." *Nature News,* January 23, 2015. https://www.nature.com/news/genghis-khan-s -genetic-legacy-has-competition-1.16767.

Dawkins, Richard, and John Richard Krebs. "Arms Races between and within Species." *Proceedings of the Royal Society B* 205 (1979): 489–511.

Finkelman, Fred D. "The Price We Pay." *Nature* 484 (2012): 459.

Foster, K. R., and H. Kokko. "The Evolution of Superstitious and Superstition-Like Behaviour." *Proceedings of the Royal Society B* 276 (2009): 31–37.

Francis, Pope. "*Laudato Si*: On Care for Our Common Home." Encyclical Letter, May 24, 2015. http://www.vatican.va/content/francesco/en /encyclicals/documents/papa-francesco_20150524_enciclica-laudato-si .html.

Gardiner, Stephen M. A *Perfect Moral Storm: The Ethical Tragedy of Climate Change*. Oxford: Oxford University Press, 2011.

"Global Warming of 1.5°C: Summary for Policymakers." IPCC Special Report, Intergovernmental Panel on Climate Change, Geneva, 2018. https://www .ipcc.ch/sr15/chapter/spm/.

"Greatest Number of Descendants." Guinness Book of World Records, n.d. http://www.guinnessworldrecords.com/world-records/67455-greatest -number-of-descendants.

Haselton, Martie G. "Error Management Theory." In *Encyclopedia of Social Psychology*, 2 vols., ed. Roy F. Baumeister and Kathleen D. Vohs, 1: 311–312. Thousand Oaks, CA: Sage, 2007.

Haselton, Martie G., and Daniel Nettle. "The Paranoid Optimist: An Integrative Evolutionary Model of Cognitive Biases." *Personality and Social Psychology Review* 10 (2006): 47–66.

Johnson, Dominic D. P. *Overconfidence and War: The Havoc and Glory of Positive Illusions*. Cambridge, MA: Harvard University Press, 2004.

Johnson, Dominic D. P., Daniel T. Blumstein, James H. Fowler, and Martie G. Haselton. "The Evolution of Error: Error Management, Cognitive Constraints, and Adaptive Decision-Making Biases." *Trends in Ecology & Evolution* 28 (2013): 474–481.

Martincorena, Iñigo, Aswin S. N. Seshasayee, and Nicholas M. Luscombe. "Evidence of Non-Random Mutation Rates Suggests an Evolutionary Risk Management Strategy." *Nature* 485 (2012): 95–98.

Milewski, Antoni V., Truman P. Young, and Derek Madden. "Thorns as Induced Defenses: Experimental Evidence." *Oecologia* 86 (1991): 70–75.

"Mother's Day: Five Incredible Records." Guinness Book of World Records, March 14, 2015, "Most Prolific Mother." http://www.guinnessworldrecords .com/news/2015/3/mother%E2%80%99s-day-five-incredible-record -breaking-mums-374460.

Nesse, Randolph M. "The Smoke Detector Principle." *Annals of the New York Academy of Sciences* 935 (2001): 75–85.

Neuhoff, John G. "Looming Sounds Are Perceived as Faster Than Receding Sounds." *Cognitive Research: Principles and Implications* 1 (2016): 15.

Oberzaucher, Elisabeth, and Karl Grammer. "The Case of Moulay Ismael—Fact or Fancy?" *PLoS One* 9 (2014): e85292.

Oreskes, Naomi, and Erik M. Conway. *Merchants of Doubt: How a Handful of Scientists Obscured the Truth on Issues from Tobacco Smoke to Global Warming.* New York: Bloomsbury, 2011.

Orrock, John L., Andy Sih, Maud C. O. Ferrari, Richard Karban, Evan L. Preisser, Michael J. Sheriff, and Jennifer S. Thaler. "Error Management in Plant Allocation to Herbivore Defense." *Trends in Ecology and Evolution* 30 (2015): 441–445.

Pascal, Blaise. *Pascal's Pensées.* Introduction by T. S. Eliot. Boston: E. P. Dutton, 1958.

Sagarin, Rafe. *Learning from the Octopus: How Secrets from Nature Can Help Us Fight Terrorist Attacks, Natural Disasters, and Disease.* New York: Basic Books, 2012.

Sagarin, Raphael. "Adapt or Die: What Charles Darwin Can Teach Tom Ridge about Homeland Security." *Foreign Policy,* September / October 2009.

Sagarin, Raphael D., Candice S. Alcorta, Scott Atran, Daniel T. Blumstein, Gregory P. Dietl, Michael E. Hochberg, Dominic D. P. Johnson, et al. "Decentralize, Adapt and Cooperate." *Nature* 465 (2010): 292–293.

Sagarin, Raphael D., and Terence Taylor. *Natural Security: A Darwinian Approach to a Dangerous World.* Berkeley: University of California Press, 2008.

Trivers, Robert. *The Folly of Fools: The Logic of Deceit and Self-Deception in Human Life.* New York: Basic Books, 2011.

Wiley, R. Haven. "Errors, Exaggeration, and Deception in Animal Communication." In *Behavioral Mechanisms in Evolutionary Ecology,* ed. Leslie A. Real, 157–189. Chicago: University of Chicago Press, 1994.

Young, Truman P. "Herbivory Induces Increased Thorn Length in *Acacia drepanolobium.*" *Oecologia* 71 (1987): 436–438.

Young, Truman P., and Bell D. Okello. "Relaxation of an Induced Defense after Exclusion of Herbivores: Spines on *Acacia drepanolobium.*" *Oecologia* 115 (1998): 508–513.

11. Our Inner Marmot

Blumstein, Daniel T. "Fourteen Lessons from Anti-Predator Behavior." In *Natural Security: A Darwinian Approach to a Dangerous World,* ed. Raphael D. Sagarin and Terence Taylor, 147–158. Berkeley: University of California Press, 2008.

Ehrlich, Paul R., and Daniel T. Blumstein. "The Great Mismatch." *BioScience* 68 (2018): 844–846.

"Facts about Skiing / Snowboarding Safety." National Ski Areas Association, Fact Sheet, October 1, 2012. https://www.nsaa.org/media/68045/NSAA-Facts -About-Skiing-Snowboarding-Safety-10-1-12.pdf.

Gardiner, Stephen M. *A Perfect Moral Storm: The Ethical Tragedy of Climate Change.* Oxford: Oxford University Press, 2011.

Ip, Greg. *Foolproof: Why Safety Can Be Dangerous and How Danger Makes Us Safe.* New York: Little, Brown, 2015.

Lin, Rong-Gong, II. "Californians Need to Be So Afraid of a Huge Earthquake That They Take Action, Scientists Say." *Los Angeles Times,* May 27, 2017.

Matter, P., W. J. Ziegler, and P. Holzach. "Skiing Accidents in the Past 15 Years." *Journal of Sports Sciences* 5 (1987): 319–326.

McMillan, Kelley. "Ski Helmet Use Isn't Reducing Brain Injuries." *New York Times,* December 31, 2013.

Mele, Christopher. "How to Get People to Evacuate? Try Fear." *New York Times,* October 6, 2016.

Page, Charles E., Dale Atkins, Lee W. Shockley, and Michael Yaron. "Avalanche Deaths in the United States: A 45-Year Analysis." *Wilderness & Environmental Medicine* 10 (1999): 146–151.

Sagarin, R., D. T. Blumstein, and G. P. Dietl. "Security, Evolution and." In *Encyclopedia of Evolutionary Biology,* 4 vols., ed. R. M. Kliman, 4: 10–15. Oxford: Academic Press, 2016.

Verini, James. "Meth Mouth: Tom Siebel's Brash Anti-Crystal Campaign." *Fast Company,* May 1, 2009. https://www.fastcompany.com/1266054/meth -mouth-tom-siebels-brash-anti-crystal-campaign.

Vrolix, Klara. "Behavioral Adaptation, Risk Compensation, Risk Homeostasis and Moral Hazard in Traffic Safety: Literature Review." Report RA-2006- 95, Universiteit Hasselt, September 2006, https://pdfs.semanticscholar.org /8070/d6cdc9dde91dbfee8f776d43a89e701e8313.pdf.

12. Wisely Living with Fear

Bernstein, Ethan S., and Stephen Turban. "The Impact of the 'Open' Workspace on Human Collaboration." *Philosophical Transactions of the Royal Society B* 373 (2018): 20170239.

Blumstein, Daniel T. *Eating Our Way to Civility: A Dinner Party Guide.* Los Angeles: Marmotophile Publishing, 2011.

Christie, Les. "'Deadliest Catch' Not So Deadly Anymore." *CNN Money,* July 27, 2012. https://money.cnn.com/2012/07/27/pf/jobs/crab-fishing-dangerous -jobs/index.htm.

"Commercial Fishing Deaths—United States, 2000–2009." *Morbidity and Mortality Weekly Report,* Centers for Disease Control, July 16, 2010. https://www.cdc.gov/mmwr/preview/mmwrhtml/mm5927a2.htm.

De Martino, Benedetto, Colin F. Camerer, and Ralph Adolphs. "Amygdala Damage Eliminates Monetary Loss Aversion." *Proceedings of the National Academy of Sciences USA* 107 (2010): 3788–3792.

Dimitroff, Stephanie J., Omid Kardan, Elizabeth A. Necka, Jean Decety, Marc G. Berman, and Greg J. Norman. "Physiological Dynamics of Stress Contagion." *Scientific Reports* 7 (2017): 6168.

Ehrlich, Paul R., and Daniel T. Blumstein. "The Great Mismatch." *BioScience* 68 (2018): 844–846.

Fischhoff, Baruch, and John Kadvany. *Risk: A Very Short Introduction.* Oxford: Oxford University Press, 2011.

Grubb, Michael A., Agnieszka Tymula, Sharon Gilaie-Dotan, Paul W. Glimcher, and Ifat Levy. "Neuroanatomy Accounts for Age-Related Changes in Risk Preferences." *Nature Communications* 7 (2016): 13822.

"Homicide Report: A Story for Every Victim." *Los Angeles Times,* continually updated. https://homicide.latimes.com/. Ishii, Akiko, Yasushi Kiyokawa, Yukari Takeuchi, and Yuji Mori. "Social Buffering Ameliorates Conditioned Fear Responses in Female Rats." *Hormones and Behavior* 81 (2016): 53–58.

ProCon.org, "International Firearm Homicide Rates: 2010–2015." ProCon.org, Santa Monica, CA, August 7, 2017. https://gun-control.procon.org/view.resource.php?resourceID=006082.

———. "US Gun Deaths by Year." ProCon.org, Santa Monica, CA, August 29, 2018. https://gun-control.procon.org/view.resource.php?resourceID=006094.

Turkle, Sherry. *Reclaiming Conversation: The Power of Talk in a Digital Age.* New York: Penguin, 2016.

van der Linden, Sander, Edward Maibach, John Cook, Anthony Leiserowitz, and Stephan Lewandowsky. "Inoculating against Misinformation." *Science* 358 (Dec. 1, 2017): 1141–1142.

"Why Is the Pain of Losing Felt Twice as Powerfully Compared to Equivalent Gains? Loss Aversion, Explained." The Decision Lab website, n.d., https://thedecisionlab.com/biases/loss-aversion/.

Acknowledgments

This book reflects lessons I've learned from over thirty years of research on antipredator behavior and fear in animals and humans. My empirical research has been generously supported by the US National Science Foundation, the Australian Research Council, the US National Institutes of Health, the German Academic Exchange Foundation, the National Geographic Society, the Paul S. Veneklasen Research Foundation, the Rocky Mountain Biological Laboratory, as well as by the University of California, Los Angeles; University of California, Davis; and Macquarie University. I was an American Institute of Pakistan Studies Fellow and a Fulbright Fellow while working in Pakistan.

My integrative knowledge about animal behavior was possible only with outstanding mentorship from Ken Armitage, David Armstrong, Walter Arnold, Marc Bekoff, Dick Coss, Chris Evans, Peter Marler, Don Owings, and especially Judy Stamps.

Over the years I have learned a lot about antipredator behavior and the nature of fear from many discussions and correspondence with Louise Barrett, Bill Bateman, Peter Bednekoff, Guy Beauchamp, Joel Berger, Oded Berger-Tal, Carl Bergstrom, Eduardo Bessa, Amos Bouskila, Tim Caro, Alex Carthey, Colleen Cassady St. Clair, Doug Chivers, Mathias Clasen, Michael Clinchy, Bill Cooper, Darren Croft, Eberhardt Curio, Sasha Dall, Chris Dickman, Larry Dill, John Endler, Esteban Fernández-Juricic, Maud Ferrari, Patricia Flemming, Benjamin Geffroy, Alison Greggor, Andrea Griffin, Rob Harcourt, James

Hare, John Hoogland, Dominic Johnson, Menna Jones, Matt Kauffman, Peter Kaye, Arik Kershenbaum, Mike Letnic, Steve Lima, Marc Mangel, Marta Manser, Dean Mobbs, Anders Møller, Raquel Monclús, Katherine Moseby, Michael Parsons, Bree Putman, Rafe Sagarin, Wendy Saltzman, Diogo Samia, Roger Seyfarth, Andy Sih, Jenn Smith, Ted Stankowich, Terrence Taylor, Dirk Van Vuren, Geerat Vermeij, Rebecca West, Ron Ydenberg, and Liana Zanette.

Although I collaborate with people around the world, I find myself particularly lucky to have colleagues at my home institution with whom I have bounced around various ideas about the nature of fear that include Mike Alfaro, Paul Barber, Aaron Blaisdell, Greg Bryant, Steve Cole, Dan Fessler, Peggy Fong, Patty Gowaty, Greg Grether, Martie Haselton, Barbara Natterson-Horowitz, Jamie Lloyd-Smith, Peter Nonacs, Noa Pinter-Wollman, Mason Porter, Tom Smith, Blaire Van Valkenburgh, Bob Wayne, and Pamela Yeh.

I've been fortunate to have published papers with about 200 undergraduate students, many graduate students, and some wonderful postdoctoral fellows. I am extremely grateful for the energy and ideas they have all brought to my life. I'm thankful for the Field and Marine Biology Quarter program at UCLA Department of Ecology and Evolutionary Biology and the Rocky Mountain Biological Laboratory's education program for financially supporting many undergraduate projects.

While based at UCLA I'm fortunate to have been able to spend between two and four months a year conducting fieldwork in Colorado where, in addition to studying marmots, birds, and deer, I've learned a lot over happy hours and dinner parties from my friends and colleagues at the Rocky Mountain Biological Laboratory. Among those, Ian Billick and Jennie Reithel, Anne and Paul Ehrlich, and John and Mel Harte stand out—they taught me *a lot* about environmental policy. Countless discussions over dinner parties in Los Angeles with our good friends, including Jeanie and Paul Barber, Charlie Saylan and Maddalena Bearzi, Eugene Volkh and Leslie Pireira, Cully Nordby and Mark Frye, David Paige and Sorel Fitzgibbon, David Neelen and Kerry Jensen, as well as Mark Gold, Rob Harcourt, Dominic Johnson, Peter

Karieva, Mike Letnic, Rafe Sagarin, and Tom Smith have helped shape how I think about the science-evidence-policy interface.

Surfing with Paul Barber in Los Angeles and Rob Harcourt in Sydney kept me grounded (or at least wet) during writing bouts.

My friends Maddalena Bearzi, Kathryn Bowers, Doug Emlen, and Barbara Natterson-Horowitz gave me excellent advice about communicating with a nontechnical audience and gave me feedback on one or more drafts of my book proposal.

I thank UCLA for supporting the two sabbatical quarters during which I wrote the first draft of this book. Mike Letnic hosted me at the University of New South Wales, where I wrote much of the first half of the book. Two years later, Rob Harcourt hosted me at Macquarie University, where I wrote more. Oded Berger-Tal guided my application for a Rothschild Fellowship visit to Ben Gurion University of the Negev along with visits to Hebrew University and Tel-Aviv University. I gave a variety of university and public talks while on these sabbaticals, including a keynote address at the Aarhus Institute of Advanced Studies as part of an interdisciplinary workshop on fear, where I learned about a variety of other disciplines' perspectives on fear. I am very grateful to all of my hosts for their wonderful hospitality during these many visits, which helped me increase my knowledge of the nature of fear and thus helped support the development of this book.

For feedback on specific chapters or sections of the book, I am extremely grateful to Mike Letnic, Dean Mobbs, Michael Parsons, and Truman Young. Barbara Natterson-Horowitz read the entire manuscript, and her astute comments helped increase the clarity of my thoughts and prose. Five anonymous reviewers provided additional constructive comments, and I hope that by addressing them the book is stronger.

Two friends and colleagues deserve special recognition. The late Rafe Sagarin was generous with his time and ideas. I learned so much about the lessons life has to offer for other disciplines by working with Rafe and our natural security working group. We have all lost so much by losing him, for he had so much more to contribute. My integrative work in the field of evolutionary medicine with Barbara Natterson-Horowitz has been and continues to be remarkably rewarding. Barbara has brought insightful and generative conversations to many a dinner party and has been a constant cheerleader as this project developed.

I struck the jackpot with my editor at Harvard University Press, Janice Audet, who immediately believed in the project and offered guidance and comments throughout the process of making a proposal into a book. Her editorial input to the penultimate drafts was transformative. This book is so much better because of Janice's insights and I am *extremely* grateful for her guidance. Katherine Brick was an *extraordinary* copy editor who, with clarity and grace, helped unpack ideas and simplify prose. Of course, I alone am responsible for any remaining errors. In addition to Janice and Katherine, I also thank the rest of the production team at Harvard.

Finally, I would not be who I am were it not for the unconditional love and support I have received from my parents, my wife and best friend, Janice, and our son, David. Janice and David have shared a number of thrilling experiences with me over the years, and I look forward to sharing many more! And, as I only recently discovered, Janice has a hidden talent as a dog trainer who successfully employs love, not fear, to encourage Theo, our corgi, to be the best boy he can be. Love trumps fear. A fitting insight with which to end this book.

Index

anthropodenial, 5

anthropomorphism, 4–5

antipredator behavior: attention and, 37, 38; escape response, 36; of fish, 7; of golden marmots, 10; group-size effects, 34; human evolution and, 3–4; human *vs.* nonhuman, 4; location of exposure to predators and, 33–34; multipredator hypothesis, 34–35; study of, 4, 129

anxiety, 6, 16, 82

arachnophobia, 28–29

Arid Recovery fence line: ecological impact of, 153–54

Armitage, Ken, 139

Ashe, Dan, 144

aspen: distribution of, 146–47; foraging behavior, 145; wolf population and, 145

Atkins, Justine, 150

attention, 36–37, 38

Australia: Arid Recovery project, 152–53; extinction of native mammals, 66, 151–52; introduction of nonnative animals in, 65–66, 151

Australian swamp rats: predators of, 67

autoimmune diseases, 130

Baghdad's Green Zone, 81

Bahari, Maziar, 102

Barrett, Louise, 124

Bayes, Thomas, 101

Bayesian logic of decision making, 101–2

Bednekoff, Peter, 74

behavior: adaptive, 84, 85, 95; changes in, 100; of closely related species, 87; diversity of, 84, 87; evolutionary

fitness in, 84–85; learning influences of, 100; shaking and hiding, 4; shared, 7. *See also* animal behavior; escape behavior

behavioral ecology, 87, 88

Benyus, Janine, 129

benzodiazepines, 11

Berger, Joel, 79–80, 146, 148

Bernstein, Ethan, 196

Beschta, Robert, 144, 145, 146, 147

Bessa, Eduardo, 93

biomimicry, 129

birds: airplane strike statistics, 38; antipredator behavior of, 83–84; of the Atherton Tablelands, 83; effect of predator sound on, 43; escape behavior of, 88–90, 91, 92; flight-initiation distance, 89–91, 110–11; reaction to added noise, 51; response to approaching object, 39–40; in Southern California, 111; tolerance of people, 93, 111

Blaisdell, Aaron, 37

blood chemicals, 63

body size: effect of, 89, 90, 91, 92, 137

Boeing 737 Max accidents, 191

Boonstra, Rudy, 18

Bowers, Kathryn, 6, 19

Bowling Alone (Putnam), 155

Boyup Brook, town of, 56

"Boy Who Cried Wolf" (Aesop), 108, 132

Brackman Island, 41, 42

brain: biochemical reaction in, 13; fear circuits, 15, 16; neural activity in, 16; physiology of, 14; response to fear, 13, 16, 21–22; structural changes in, 199; ventromedial prefrontal

"flush early and avoid the rush" hypothesis, 90

flycatchers, 25

Folly of Fools, The (Trivers), 164

Foolproof: Why Safety Can be Dangerous and How Danger Makes Us Safe (Ip), 177

functional magnetic resonance imaging (fMRI), 15–16, 54

Garcetti, Eric, 192

Gardiner, Steve, 182, 183

Geffroy, Benjamin, 93

Genghis Khan, 163

Giving Up Density (GUD) technique, 73–74

glucocorticoids, 16

golden marmots: antipredator behavior of, 10

Gorongosa National Park, 150, 182

grackles: response to noise, 51

Gray, George, 32

greater bilbies, 153, 154

greenhouse gases: break down of atmospheric, 183–84

Green Mile, The (film), 51

Griffin, Andrea, 31, 33, 81, 102, 106, 153

ground squirrels, 75; alarm calls of, 117, 192

gun violence, 191, 192

habituation: body size and, 112; delay of, 174–75; mechanisms of, 108, 109; natural history of, 110; prediction of conditions of, 181

habituation-recovery protocol, 132

hares: population cycle, 18; stress hormone levels, 18–19

harlequin filefish, 64

Haselton, Martie, 160, 161

Henzi, Peter, 124

hermit crabs: impact of human noises on, 37–38; studies of, 36, 37

Herodotus: account on fantastic animals, 69

Hettena, Alex, 44, 45

Hitler, Adolf, 164

Hockett, John, 7

How Monkeys See the World (Cheney and Seyfarth), 123

HPA axis, 15, 17

human fear: as agent of change, 174, 175; aging and, 199; as cause of war, 175; disgust as motivator of, 172; of forest, 78–79; of HIV, 192; of homeowners, 156; of horrific outcomes, 190–91; of hypothermia, 41–42; of infection, 162; illogical nature of, 186–87; of loss, 189; manipulation with, 197–98; of nuclear weapons, 175; of outsiders, 181; politics and, 178–79, 181, 182; of predators, 35–36, 144, 157; of punishment, 194; of sharks, 9; of snakes, 28; social effect of, 9, 196; socially transmitted, 106, 107; of spiders, 28–29; survival and, 181; of terrorist attacks, 174; tribalism and, 155–56; of the unknown, 191–92; of violence, 155

humans: in Australia, introduction of, 151; behavioral changes, 183, 198; communication abilities, 7, 196; decision-making strategies, 89; ecological consequences of fear in, 154–55; effect of adrenaline on,

humans (*continued*)
13–14; emotional responses in, 5, 54, 61; evolutionary history of, 78; facial expressions of, 54; homicide rates, 188; impact of chronic stress on, 43; innate abilities of, 30; learning abilities of, 180–81; life-history theory of, 96; loss-aversion bias, 189; noises of, 37–38; perceptions of safety, 78; response to scents, 60, 61, 67–68; response to threats, 179–80, 198–99; risk assessment, 9, 81–82, 187; risks and benefits management, 95–96; scents of, 59; self-delusions of, 164; sensitivity to misinformation, 134–35, 198; superstitious behaviors of, 166–67

human shield effect, 93, 94

Hunter, Duncan, 107

Hussein, Saddam, 175

hypothalamus, 14

hypothermia, 41

hysterical contagion, 106

ibis: escape behavior of, 93

iguanas: escape behavior of, 27

immune system, 129

inducible defenses, 167

intraguild predation, 152

Ip, Greg, 177

Iraq, 81–82, 107, 175

island tameness, 84

Jacob, François, 179

Japanese macaques: fear of snakes, 28

Johnson, Dominic, 159, 160, 164, 165

Johnson, Lyndon B.: *Daisy* advertisement, 55, 178

Kakamega National Park, 1

Kalahari Meerkat Project, 128

Kangaroo Island, 31

kangaroos: behavioral study of, 56–57; effect of predator smell on, 57–58; emergence from cover, 72; fear of dingoes, 66; impact of mining operations on, 56; repellents for, 56; response to alarm calls, 72; size of, 72

Karakoram Mountains, 10–11

Kauffman, Matt, 144, 146, 147, 148

Kaye, Peter, 51, 52

Kennedy, John F., 165

Khrushchev, Nikita, 165

Khunjerab National Park, 10

Kim Jong Un, 175

Krebs, Charlie, 18

Laubach, Zach, 50

"*Laudato Si:* On Care for Our Common Home" encyclical letter, 169

learning: from bad experiences, 100; Bayesian logic of, 104, 114; cost of, 103; definition of, 100; effect on risk assessment, 197; location-specific and context-specific, 102–3; rapid, 100–101, 102; social, 106; transfer across similar sorts of predators, 105–6

Learning from the Octopus (Sagarin), 159

Leigh, Janet, 50

Letnic, Mike, 151

life-dinner principle, 166

lionfish: elimination in the Caribbean, 97

lizard escape behavior, 92

location, 193–94; ecological perspective on, 197; economic logic of, 96, 97; effect of learning on, 197; fifteen principles of, 188–97; loss-aversion bias and, 189; pre-existing beliefs and, 188–89; scope of damage and, 193; statistical methods of, 195; studies of, 47, 76, 161–62; of subordinate and dominant animals, 75–76; type of outcome and, 190–91. *See also* trade-offs, starvation-predation risk

risk compensation, 176, 178

risk taking: balance between safety and, 186

rock wallabies: antipredator behavior in, 83

Rocky Mountain Biological Laboratory, 66, 139–40, 141

rodents: alarm signaling of, 127; foot drumming, 135

Rwandan genocide, 107

safety: public perception of, 177–78; risk compensation and, 178

Sagan, Carl, 28

Sagarin, Raphael (Rafe), 158, 159

Salem witch hunt, 106

Samia, Diogo, 93

Sapolsky, Robert, 155

scent: chemicals associated with, 63–64; of danger, 67, 68; fear and, 57; of predators, 60–61

Schmich, Mary, 200

scientific progress: idea of, 6

screams, 50

sea turtles: population of, 79

sensitization, 110, 181

sexual attraction, 160, 163

Seyfarth, Robert, 122, 123, 125, 132

Shelley, Erin, 127

Shubin, Neil, 5–6

signal detection theory. *See* error management theory

Skiles, Jeffrey, 38

Smith, Doug, 144

smoke detector principle, 163, 194

snakes: fear of, 27–28

social stressors, 3, 5, 17

sound: associated with increased risk, 135–36; distortion of, 49; experimental study of, 42–43; of fear, 53, 55; frequency change of, 49; noisy and nonlinear, 50; of predators, 43–43, 44, 45; rough, 54–55. *See also* noise

sparrows: song, behavior of, 42; white-crowned, response to noise, 51; white-throated, foraging behavior of, 75–76

spiders: fear of, 28–29

squirrels, 75; alarm vocalizations of, 117; foraging habits of, 74; response to alarm calls, 137

starlings: alarm calls, 137–38

starvation-predation risk trade-offs. *See* trade-offs, starvation-predation risk

stoats: foraging behavior of, 66–67

Streetcar Named Desire, A (film), 50

stress, 16, 17, 75

Struhsaker, Thomas, 122

Sullenberger, Chesley, 38

Suraci, Justin, 148

surfing: competitions, 95; joys and danger of, 98–99

TAAR4 receptors, 62–63, 68

tammar wallabies: antipredator behavior of, 33, 34; emergence from cover, 72; exposure to predators, 33; of Kangaroo Island, 33–34; learning abilities of, 81, 102, 106; New Zealand population of, 32–33; predator recognition template of, 33, 35, 103; response to alarm, 72, 81; size of, 72; South Australian population, 32; susceptibility to capture myopathy, 19–20

Tanganyika laughter epidemic, 106

Tasmanian animals, 58, 67, 151

Taylor, Terence (Terry), 159

templates, 30, 31. *See also* predator recognition templates

terrorist attacks, 130, 174, 198

threat: biochemical responses to, 21–22

Toxoplasma gondii, 65

trace amine-associated receptors (TAARs), 62

trade-offs, starvation-predation risk, 8, 18, 73, 80–82, 130, 161

*trans-*4, 5-epoxy-(E)-2-decenal (E2D), 63, 67

tree frogs: accelerated hatching in, 103

trees: fear associated with, 78–79

trimethylthiazoline (TMT), 61, 67

Trivers, Robert, 164

trophic cascades: creation of, 143, 149–51; environments of, 151; example of, 145–46; fear-based, 144, 148–49

Trump, Donald, 107–8, 175, 184

Turban, Stephen, 196

Turkle, Sherry, 196

2-phenylethylamine (PEA), 62, 67

United Nations Intergovernmental Panel on Climate Change, 170

United States: deaths statistics, 130, 191–92; homicide rate, 188; occupation of Iraq, 81–82; war in Afghanistan, 130

urban prey, 93

urine, chemical profile of, 58–59; of predators, 56–57, 58, 59, 62, 63; of prey, 60, 63. *See also* dingo urine, dog urine, mouse urine

US Airways Flight 1549 accident, 38

"Us versus Them" categorizations, 155

Valium: medical effect of, 11–12

Vancouver Island, 141, 148, 149

vervet monkeys: alarm calls of, 123, 124, 125; behavior changes of, 123; foraging habits of, 123; referential abilities of, 124–25; response to predators, 123–24; response to starling alarm calls, 137–38; studies of, 122–23

Virgin Islands: hermit crabs' population, 37

vocalization: communication through, 116; and fear, 25; individually distinctive, 131

vocal production systems, 49–50

Voyage of the Beagle (Darwin), 84

Warkentin, Karen, 103

war paradox, 164–65

Waser, Nick, 141, 142

Wegener, Alfred, 77

wolves: ecological impact of, 145; growing population of, 143; hunting habits of, 148; reintroduction to the